Y0-DCC-109

Dissection of the cat

AND COMPARISONS WITH MAN

A laboratory manual on *Felis domestica*

MOSBY

␣ROR

For prompt service, call (314) 872-83␣

Instructor's C

SUGG␣␣␣ ␣ST R␣␣

Dissection of the cat

AND COMPARISONS WITH MAN

A laboratory manual on *Felis domestica*

BRUCE M. HARRISON

Late Emeritus Professor of Zoology
The University of Southern California
Los Angeles, California

REVISION CONSULTANT
L. E. St. Clair
College of Veterinary Medicine, University of Illinois
Urbana, Illinois

Drawings by Scientific Illustrators
Champaign, Illinois
ARTIST
Carol Arian

with 72 illustrations

Seventh edition

The C. V. Mosby Company

St. Louis   1976

Seventh edition

Copyright © 1976 by The C. V. Mosby Company

All rights reserved. No part of this book may be reproduced in any manner
without written permission of the publisher.

Previous editions copyrighted 1948, 1952, 1956, 1962, 1966, 1970

Printed in the United States of America

International Standard Book Number 0-8016-2075-9

Distributed in Great Britain by Henry Kimpton, London

CB/CB/B   9   8   7   6   5   4   3   2   1

Preface

This dissection atlas has been widely used over the years in at least two kinds of courses: those in human anatomy when the human body is not available for dissection by the student and those in comparative anatomy in which the thorough study of the cat is a primary goal. We hope this revision will meet the needs of students in these courses more than ever before.

Based on much feedback from users, this edition includes revisions of two important aspects, the illustrations and the terminology. Every pedagogically important view of the cat's body has been drawn from newly dissected material. The result is a complete new set of accurate and esthetically pleasing artwork. Through-out the book, terminology has been changed and updated to conform to the current standards of *Nomina Anatomica* for man and *Nomina Anatomica Veterinaria* for the cat. These changes were made in order to keep all the terminology consistent and uniform and to promote carryover from the anatomy of the cat to the anatomy of man.

Dr. L. E. St. Clair, College of Veterinary Medicine, University of Illinois, did the new dissections, completely supervised the preparation of the new artwork, and revised the terminology. He also revised portions of the manuscript itself to increase the clarity of dissection instructions.

The Publisher

Contents

Introduction, 1

ONE

Skeleton of the cat, 2

Skull, dorsal, 2
Sutures of skull, 6
Skull, ventral, 6
Skull, lateral, 7
Skull, median sagittal, 8
Development of skull, 9
Foramina of skull, 9
Some differences in skulls of the cat and man, 10
Mandible, 13
Pectoral girdle, 14
Humerus, 14
Radius and ulna, 15
Spinal column (cervical, thoracic, lumbar, sacral, and coccygeal vertebrae), 16
Cervical vertebrae, 16
Thoracic vertebrae and ribs, 17
Lumbar vertebrae, 18
Sacral vertebrae, 19
Coccygeal vertebrae, 19
Some differences in vertebral, or spinal, columns of the cat and man, 20
Pelvic girdle, 25
Femur, 25
Tibia and fibula, 26
Growth of long bones, 26
Some differences in skeletons of the cat and man, 26

TWO

Muscles of the cat, 31

Four major groups of skeletal muscles, 32
Integumentary, or cutaneous, muscles, 32
Superficial muscles of neck and shoulder, 33
Deep muscles of neck and shoulder, 35
Ventral view of pectoral muscles, 37
Caudal muscles of brachium, or upper arm, 43

Medial muscles of shoulder and cranial muscles of upper arm, 44
Superficial lateral muscles of forearm, 46
Deep lateral muscles of forearm, 47
Superficial medial muscles of forearm, 48
Deep medial muscles of forearm (flexor digitorum profundus group), 51
Ventral muscles of neck, lower jaw, and thorax, 56
Branchial muscles, 57
Veins, lymph nodes, and salivary glands, 57
Muscles of thoracic and abdominal wall, 61
Superficial muscles of hip and thigh, 67
Deep muscles of hip and thigh, 69
Superficial and deeper muscles of thigh, 71
Femoral triangle of thigh, 73
Deep muscles of cranial portion of thigh, 73
Muscles of leg, 77
Deep cranial muscles of leg, 78
Deep caudal muscles of leg, 78
Some differences in muscles of the cat and man, 83
General conclusions on muscles, 83

THREE

General internal organs of the cat, 84

Survey of internal organs, 84
Thoracic organs, 90
Some differences in internal organs of the cat and man, 91
General urogenital system, 95
Female urogenital system, 95
Pregnant cat, 97
Some comparisons regarding gestation and birth, 99
Male urogenital system, 99
Development of urogenital system, 100
Homologies in urogenital systems of the cat and man, 101
Some differences in urogenital systems of the cat and man, 101

FOUR

Venous and lymphatic systems of the cat, 107

Tributaries of cranial vena cava, or precava, 107
Tributaries of brachiocephalic vein, 107
Tributaries of axillary vein, 109
Tributaries of external jugular vein, 109
Tributaries of caudal vena cava, or postcava, 110
Tributaries of external and internal iliac vein, 110
Hepatic portal system, 111
Comparisons of venous systems in the cat and man, 113
Lymphatic system, 117

FIVE

Arterial system and heart of the cat, 121

Arteries with bases within thoracic cavity, 121
Arteries to upper part of neck and head, 123
Arteries to lower part of neck and limbs, 124
Arteries of abdomen and limbs, 124
Relationships of heart, 126
Dissection of heart, 126
Coronary circulation, 127
Stages of development, 128
Some differences in arteries of the cat and man, 128

SIX

Regions of the head of the cat, 131

Some differences in the cat and man, 133
Dissection of ear, 137
Structural development of ear, 138

SEVEN

Brain of the cat, 143

Brain, median sagittal, 143
Brain, dorsal, 145
Brain, ventral, 147
Arterial supply to brain, 147
Some differences in brains of the cat and man, 148

EIGHT

Spinal cord and peripheral nerves of the cat, 151

Nerves of brachial plexus, 151
Autonomic nervous system, 152
Nerves of lumbosacral plexus, 153
Two main divisions of autonomic nervous system, 154
Spinal cord and spinal nerves, 155
Some differences in plexuses and nerves of the cat and man, 157

NINE

General summary, 161

TEN

Definitions of terms, 165

Introduction

The terms used in this book follow, in general, those listed in *Nomina Anatomica* for man and *Nomina Anatomica Veterinaria* for the cat. Some directional terms do not apply equally to man and the cat, since one stands upright and the other walks on four limbs. **Anterior** in man is **ventral** in the cat; **posterior** is **dorsal**. In the cat, the term **cranial** refers to surfaces toward the head, and **caudal** to those toward the tail. However, for surfaces on the head of the cat, the term cranial is replaced by **rostral;** thus **rostral** and **caudal** indicate opposing directions on the head. Unlike that of the cat, man's head curves and is not considered to be a prolongation of the body; thus rostral is **anterior** and caudal is **posterior.** The terms **superior** and **inferior** are used for directions toward and away from the head of man.

ONE
Skeleton of the cat

A box containing about twenty of the larger representative cat bones should have been issued to you. Select the following bones from your set and find their proper position on the drawing of the skeleton (Fig. 1-1). The first two vertebrae are the **altas** and **axis,** respectively. The **thoracic vertebrae,** to which the **true** and **false ribs** are attached, are situated in the chest. The first nine ribs are called **true ribs,** since each has its own **cartilage** attaching it to the sternum. The last four ribs are called **false ribs,** since each does not have its own attachment to the sternum. The last of the false ribs is also called a **floating rib,** since it has no cartilage attaching it to the sternum. The lower ends of the ribs join the sternum; the cranial bone of the sternum is the **manubrium,** followed by several segments constituting the **body,** or **sternebrae bones,** and the caudal portion, the **xiphoid process.** Next in the spinal column are the **lumbar vertebrae** in the small of the back, then the **sacrum,** consisting of three vertebrae, to which the **ilium** is attached. The tail bones are the **caudal (coccygeal) vertebrae.** There are usually a few remnants of hemal arches, known as **chevron bones,** on the lower surface of the fourth, fifth, and sixth caudal vertebrae.

The **clavicle** is a small bone cranial to the lower end of the **scapula,** and the **hyoid** is caudal to the lower jaw. These bones may be seen on a well-mounted skeleton. The bones of the thoracic limb consist of the following: **humerus, radius, ulna, carpals** (wrist bones), **metacarpals** (palm-of-the-hand bones), and **phalanges** (finger or toe bones).

The bones of the pelvic limb are as follows: **femur, patella** (kneecap), **tibia, fibula, tarsals** (ankle and heel bones), **metatarsals** (instep bones), and **phalanges** (toe bones).

Arrange the skull and the representative bones of the spinal column in a straight line with the head farthest away. Determine whether each of the remaining bones belongs to the right or left side, and place it in its proper position on the right or the left of the vertebrae representing the spinal column. Refer to the mounted skeleton for aid in determining whether or not a given bone is on the right or the left.

Compare Figs. 1-1 and 1-2, and observe that the skeletons of the cat and man are constructed on the same general plan. If you place the mounted cat's skeleton up on its hind legs or place the mounted human skeleton on its hands and knees and compare them, the close similarity of structure will be much more evident and striking.

Whenever you come to a scientific or technical term and you are not sure of its meaning, look in the back of the book for its definition.

SKULL, DORSAL (Fig. 1-3)

Look in the back of the manual for definitions of many of the technical terms used.

Observe the **incisive (premaxillary) bones** at the sides of the **external nasal apertures.** These bones bear the **incisor teeth.** How many are there? At each side of the middorsal line, immediately caudal to the nasal apertures, are the **nasal bones.** Lateral to these and also to the incisive bones are the **maxillary bones.** Each maxillary normally bears one **canine,** three **premolars,** and one small **molar.** Farther caudal on the dorsal surface are the **frontal bones,** which meet one another along the **middorsal suture** and project laterally as the **postorbital,** or **zygomatic, process** of the frontal bone.

A shorter discussion of the anatomy of the cat is given in Harrison, B. H.: *Manual of comparative anatomy,* ed. 3, St. Louis, 1970, The C. V. Mosby Co.

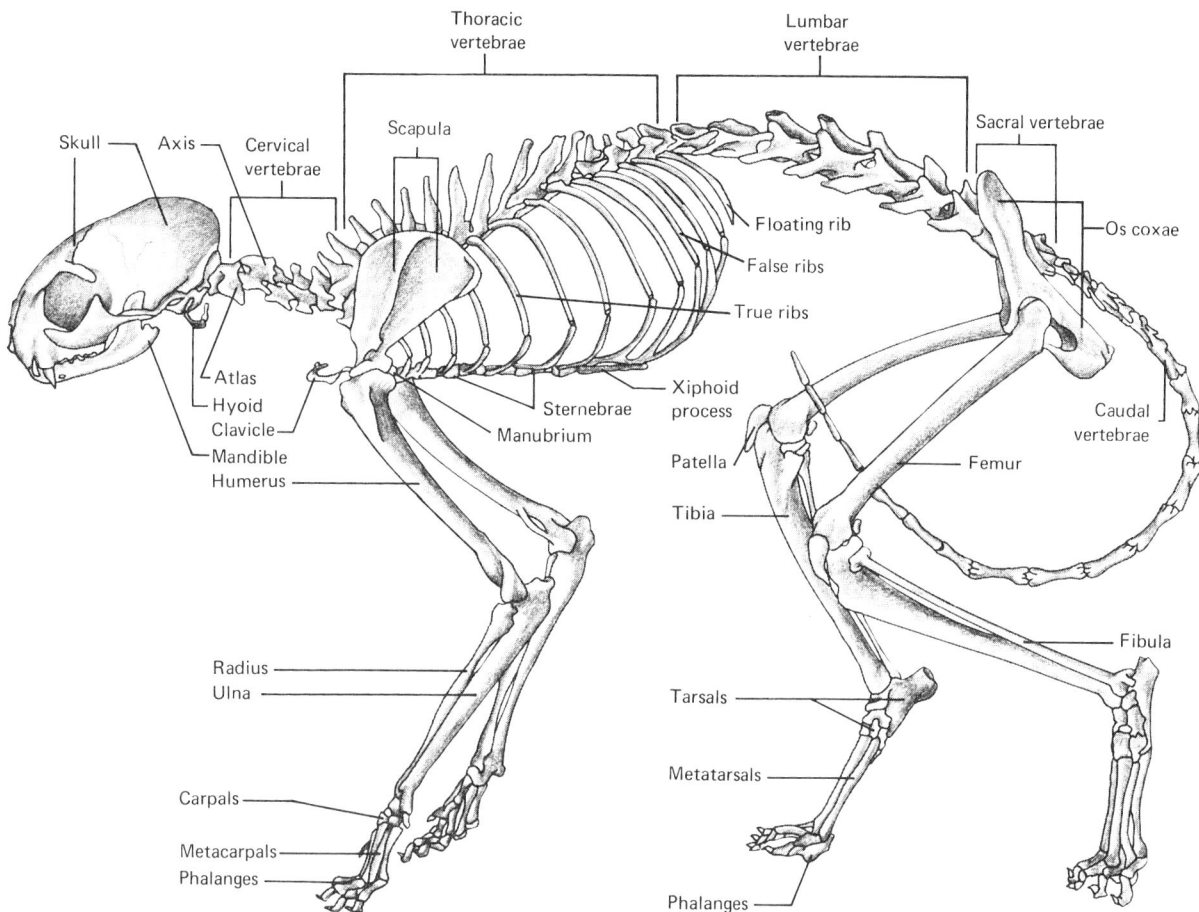

Fig. 1-1. Skeleton.

The orbit of the eye is bounded laterally by the **zygomatic arch**, which bears at its caudal extremity the **postorbital**, or **frontal, process** of the malar, directed toward the postorbital process of the **frontal bone**, mentioned in the previous paragraph. The **zygomatic arch**, joins caudally with the **zygomatic process** of the **temporal bone** to form the **zygomatic arch**. The **orbit of the eye** and the **temporal fossa** are bounded laterally by the **zygomatic arch**.

Caudal to each frontal bone, meeting one another along the middorsal line, is a **parietal bone**, and lateral to each of these is the **squamosal portion of the temporal bone**. Caudal to and slightly between the parietals sometimes may be seen a small **interparietal bone**, and caudal to it is the **supraorbital bone**. This latter bone bears a transverse ridge, known as the **nuchal crest**, for the attachment of the muscles of the dorsal part of the neck, and it also bears

a **dorsal median sagittal crest**, which extends forward between the parietals.

Within the limits of the orbit of the eye may be seen, when viewed from above, a portion of the **palatine bone** with two small openings, the palatine foramina. The more lateral opening is for the **passage of the palatine branch of the facial nerve**, and the median opening is for the **sphenopalatine branch** of the same nerve. These foramina can be seen better in ventral view. Rostral to the **palatine bone** is the **lacrimal bone**, which also bears an opening at its medial border, the **lacrimal canal**, through which pass the tears or secretion of the lacrimal gland into the nasal chamber. The part of the maxillary bone rostral to and below the front of the eye contains the large **infraorbital foramen** for the passage of the infraorbital nerves and blood vessels. These nerves are branches of the maxillary branch of the **fifth (trigeminal)**

3

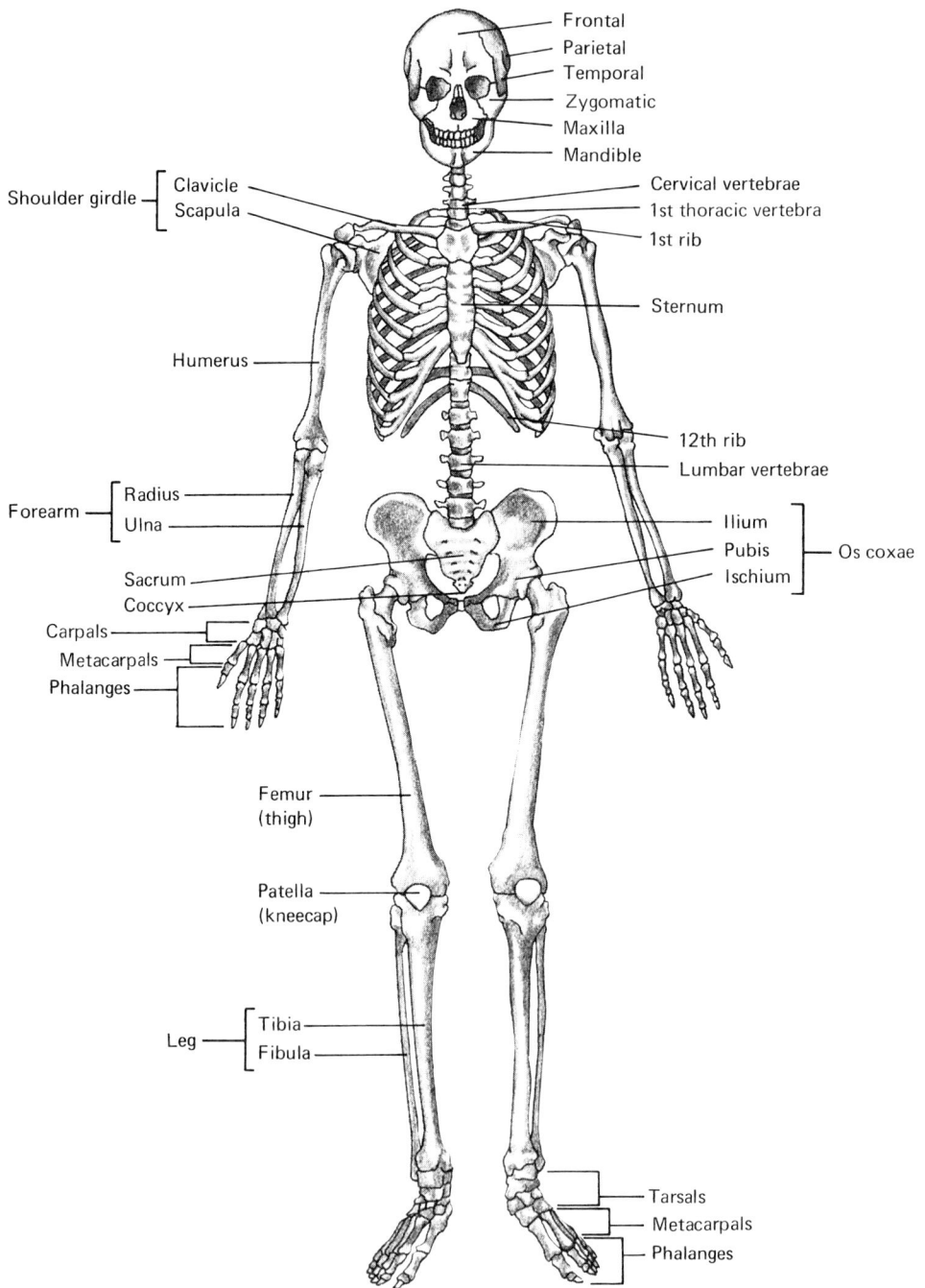

Fig. 1-2. Human skeleton, anterior view.

Frontal
Parietal
Temporal
Zygomatic
Maxilla
Mandible

Shoulder girdle
Clavicle
Scapula

Cervical vertebrae
1st thoracic vertebra
1st rib

Sternum

Humerus

12th rib
Lumbar vertebrae

Forearm
Radius
Ulna

Ilium
Pubis
Ischium

Os coxae

Sacrum
Coccyx

Carpals
Metacarpals
Phalanges

Femur
(thigh)

Patella
(kneecap)

Leg
Tibia
Fibula

Tarsals
Metacarpals
Phalanges

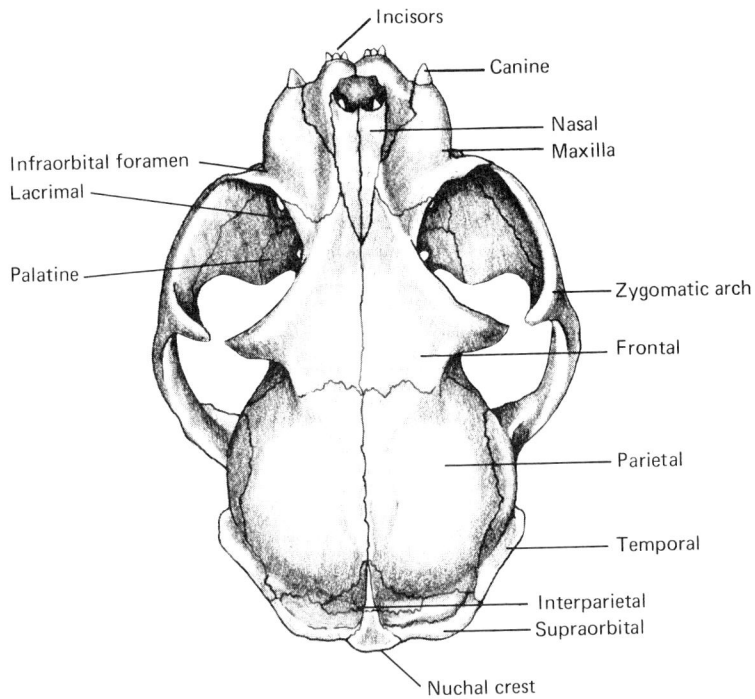

Fig. 1-3. Bones of skull, dorsal view.

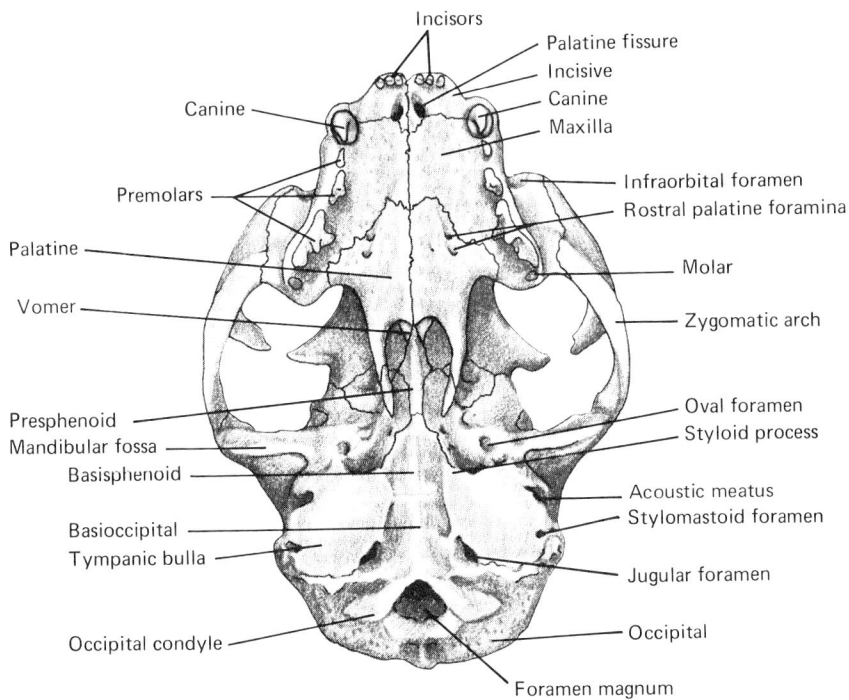

Fig. 1-4. Bones of skull, ventral view.

nerve. Other bones of the orbit will be considered later.

SUTURES OF SKULL

The bones of the skull are joined by means of immovable articulations, known as **sutures.** These sutures are designated by combining names of the bones between which they are situated, such as **nasomaxillary suture** between the nasal and maxillary bones. Often, when a suture separates the two corresponding bones of opposite sides, the prefix "inter" is used, such as the **intermaxillary suture** between the maxillaries, which shows on the ventral surface. The sutures bounding the parietal bone, however, have not been named by this system. The suture caudal to or behind the parietals, separating them from the occipital, is known as the **lambdoid suture.** Separating the parietals and squamous portions of the temporal bones is the **squamous suture,** and the transverse suture separating the two parietals from the two frontals is the **coronal suture.** Separating the two nasals, the two frontals, and the two parietals is the **median sagittal suture.**

SKULL, VENTRAL (Fig. 1-4)

As you make this study, it would be most helpful to have the disarticulated, or individual, bones of the skull to examine also.

Examine the ventral surface of a skull. The **incisive bone** almost surrounds the palatine fissure. The limiting sutures may be difficult to determine in an adult skull. The **incisive bones** bear the **incisor teeth.** The **maxillary bones** are immediately caudal, meeting one another along the midventral line, and in the fully developed specimen each bears one **canine,** three **premolars,** and one small **molar.** The **palatine bones** are caudal to the **maxillary bones,** and parts of them form the roof of the mouth, known as the hard palate. Sometimes, in the embryonic stage of man, the palatine process from one side does not come across and meet its mate from the other side, resulting in a **cleft palate.** So far as is known, this does not occur in the cat. The **vomer** lies dorsal to the median line of the maxillary and palatine bones and helps separate the two **nasal chambers;** in man the vomer pushes up into the cranial chamber, separating parts of the **cerebral hemisphere** of the **cerebrum.** The flattened part of the vomer may be seen by looking into the caudal nasal open-

ings. The soft palate, which is membranous, lies caudal to the bony part and separates the cranial part of the pharynx into the **nasopharynx** and the **stomodeal part.** The **presphenoid bone** may be seen as a small, elongated bone along the median line caudal to the vomer, forming part of the roof of the pharynx and nasal passage. Lateral to the presphenoid bone is the **pterygoid process** (see Fig. 1-5) of the sphenoid (see Fig. 1-6) bone, also forming part of the roof and sides of the pharynx and extending background and downward as long, slender projections about one-half inch apart. Each sharp point is a **hamular process** (see Fig. 1-6) of the pterygoid. Within the caudomedial wall of each orbit lie two distinct portions of the sphenoid region, which may be seen if a fairly young skull is examined. There are found four or five foramina, which include the **optic foramen,** for the passage of the optic nerve.

Ventral to the optic foramen is the **orbital fissure,** or **foramen,** through which pass the **third** (motor occulus), the **fourth** (pathetic or trochlear), and the **sixth** (abducent) cranial nerves, and the ophthalmic branch of the **fifth** (trigeminal) cranial nerve. The **optic** and **orbital foramina** may not show in a strict ventral view (see lateral view, Fig. 1-5). The greater wing of the basisphenoid bone projects dorsally and laterally as a long process between the lateral edges of the frontal and temporal bones and joins a ventrally directed process from the parietal.

In the cat, the **presphenoid** and its orbital extentions are fused into one bone, while the **basisphenoid** and its wings are fused with one another. In man, all six bones are fused into one and are called the **sphenoid.** The fusion of bones is a sign of advancement and indicates, in this small way, that man is higher than the cat. Examine these parts (if available) on disarticulated skulls of cat and man.

The **basisphenoid** lies caudal to the presphenoid along the midventral line. About halfway from this midventral line on the basisphenoid to the articulating **mandibular fossa** for the lower jaw is the **oval foramen;** rostral and medial to it is the **round foramen.** The oval foramen is for the passage of the mandibular branch of the fifth (trigeminal) nerve, and the round foramen is for the maxillary branch of the trigeminal. The last two foramina complete the series of four mentioned previously. Lateral to the oval foramen is a transverse depression, the **mandibular fossa,** for the articulation of the lower jaw. Fit the jaw into this depression.

The large, oval prominences caudal to the mandibular fossae are the **tympanic (auditory) bullae,** which are probably for the amplification of sound. Between them is the **basioccipital bone.** The large **foramen magnum** is at the caudal end of the skull and through it passes the spinal cord. The **occipital condyles** are prominences on each side of the foramen magnum and close to it. These articulate on the first vertebra of the spinal column, the **atlas.**

Close to the inner surface of the posterior extremity of the tympanic bulla is the large **jugular foramen** for the passage of the ninth, tenth, and eleventh nerves. On the rostrolateral surface of the bulla is the external **acoustic (auditory) meatus,** which leads to the tympanic membrane. Caudal to this are one or two small **stylomastoid foramina.** The small **mastoid process** of the **temporal bone** projects forward, almost covering one small foramen. At the rostromedial angle of the tympanic bulla, projecting ventrally forward and medially, is the **styloid process.** Just rostral to this process is a **foramen for the exit of the eustachian tube** from the middle ear. Only a portion of the **ethmoid bone** can be seen in the complete skull. This part constitutes the **conchae,** or **turbinate bones,** which project into the nasal chamber (see Fig. 1-6).

SKULL, LATERAL (Fig. 1-5)

Keep in mind the parts that you will see in the lateral view that you identified in the ventral view study. The **occipital condyles** project ventrally near the posterior part of the skull. Above these condyles is the **nuchal crest,** or **ridge,** that passes toward the **acoustic meatus,** which is the large opening on the lateral surface of the **tympanic bulla.** The nuchal crest becomes continuous with the **linea temporalis,** which joins the **zygomatic process** of the **temporal** and **zygomatic arch.** On the caudal surface of the bulla is the small **jugular process** of the occipital bone, a short distance lateral to the occipital condyles. Rostral to this process, about halfway to the acoustic meatus, is the much larger **mastoid process.** One or two **stylomastoid foramina,** for the passage of branches of the seventh cranial nerve, lie rostral to the mastoid process. If the skull is not thoroughly cleaned, these small openings may not be seen. Within the opening of the auditory meatus and medial to the tympanic membrane are the small ear bones, **malleus, incus,** and **stapes** (see Fig. 6-3). These cannot be seen unless the bone is cut away, which may be done later.

The foramen for the passage of the **lacrimal duct** is at the rostrolateral edge of the **lacrimal bone,** and the **infraorbital foramen** may also be seen in the lateral view. Caudal to the lacrimal and below the frontal a part of the **palatine bone** may be observed containing two openings,

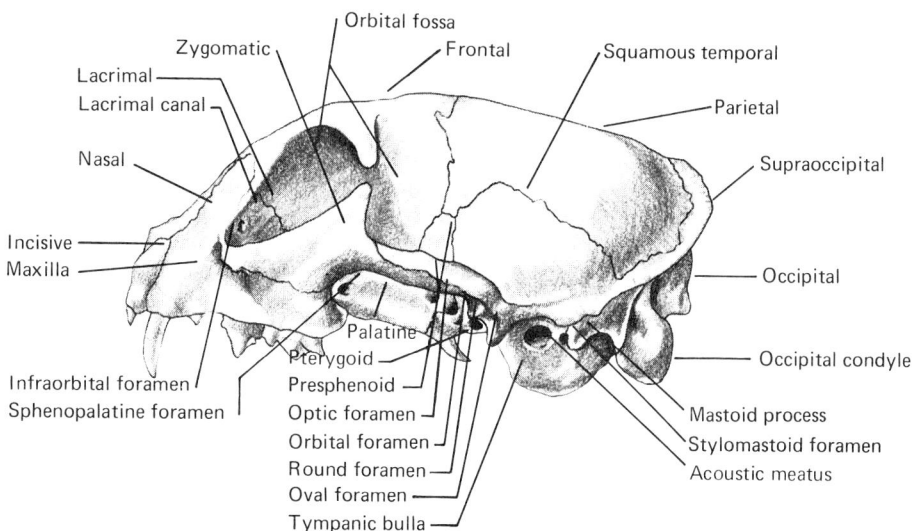

Fig. 1-5. Bones of skull, lateral view.

the outer and smaller **palatine foramen** for the passage of the **palatine branch of the maxillary nerve** and the larger and medial **sphenopalatine foramen** for the **sphenopalatine branch** of the same nerve. The former foramen is the caudal end of the palatine canal. The other end is in the roof of the mouth, in or near the suture separating the maxillary and palatine bones. A small part of the **ethmoid bone** forms a part of the rostromedial wall of the orbit between the **frontal, lacrimal,** and **palatine bones.**

SKULL, MEDIAN SAGITTAL (Fig. 1-6)

Examine the half skull. Do you have the right or the left half? The skulls that have been soaked in formalin hold the bones together much better than the fresh specimens; hence these are sawed in half for this study. We now wish to see the relationships between the bones of the skull and the larger parts of the brain. Observe that the cranial cavity is partially divided by a bony partition, or septum, which extends ventrally from the parietal and occipital bones. This is the **internal occipital protuberance (tentorium),** which is unossified in man. The larger and quite irregular cavity rostral to the **tentorium** is filled almost entirely by the **cerebrum,** with a **fossa for the olfactory bulbs,** filling the comparatively small area against the nasal chamber. The **cribriform plate,** which is a part of the ethmoid bone, separates the nasal and cranial chambers, and through it pass many olfactory nerves from the olfactory bulbs onto the **turbinate bones (conchae)** of the **nasal cavity.** There are three of

these **conchae** in man, projecting into each nasal chamber from the lateral wall. The upper two are **ethmoturbinates,** and the lower is the **maxilloturbinate.** Being much more complicated in the cat than in man, these turbinate bones have relatively more surface for nerve endings, which probably accounts in part for the keener sense of smell in the cat.

The lowest depression on the median plane of the skull, rostral to the lower end of the **tentorium,** is the **sella turcica,** or **pituitary fossa,** in which is lodged the **pituitary endocrinal gland.** Immediately dorsal to this gland is the **diencephalon** of the brain. Above the olfactory bulbs the **frontal sinus** may be seen, providing the skull has been cut slightly to one side of the median plane. This cavity is in the frontal bone. Rostral to it on each side is the **ethmoid sinus,** which is within the spaces in the ethmoid bone.

The **ethmoid sinus** is relatively larger in the cat than in man, as is also the **maxillary sinus.** It is believed that these sinuses drain better in animals that carry the body and head in a horizontal position. Man is more prone to sinus trouble. The **conchae** are projections that curve from the lateral walls into each nasal chamber. Each concha has many folds, onto which go the olfactory nerve endings for the sense of smell and many blood vessels for warming and moistening the air. The nose is an air conditioner for the lungs.

The cavity behind the **tentorium** is largely occupied by the **cerebellum** and the **medulla**

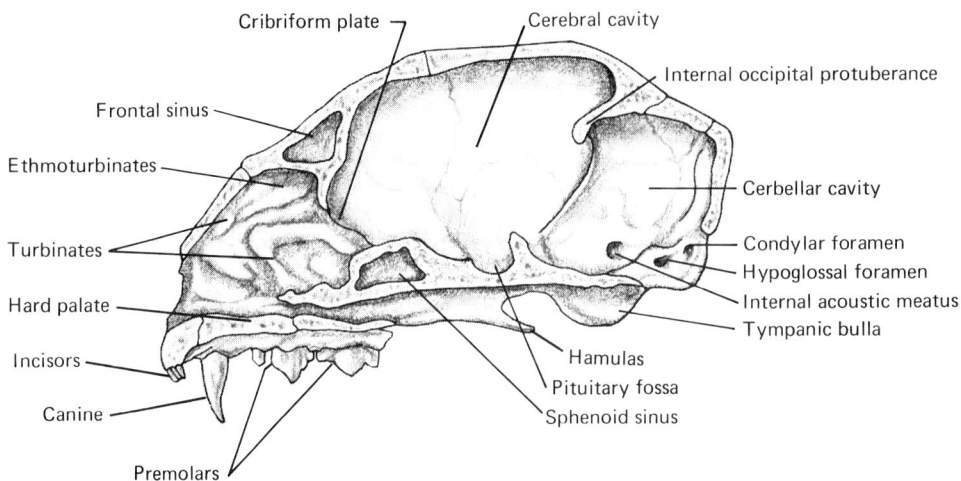

Fig. 1-6. Bones of skull, median sagittal view.

oblongata. In the lower lateral wall of the skull, which surrounds the cerebellum, is the **petrosal portion of the temporal bone,** which has an irregular inner surface and often a slightly different color. It contains the **semicircular canals** and the **cochlea.** The latter contains the **organ of Corti** of the inner ear. The inner surface of the petrous portion of the temporal bone has two depressions. Within the lower depression are two foramina. The more dorsal of the two is the **canalis facialis** for the passage of a branch of the seventh nerve, and the ventral foramen is the **internal acoustic meatus** for the passage of the eighth (auditory) nerve. The upper depression on the inner surface of the petrosal part has no foramina and is of no special significance. In the lateral wall of the **foramen magnum** are two foramina: the lower is the **hypoglossal foramen,** and the more dorsal is the **condylar foramen.** Above and rostral to the **hamular process** and the **palatine bone** is the internal naris and above it, the sphenoid sinus. There is some evidence from a comparative study of the nerves and vertebrae of various animals that the skull has been derived by a fusion of vertebrae.

DEVELOPMENT OF SKULL

There are **two principle kinds or types of bones** in the skull that differ in their embryological development.

1. The **membranous** (dermal or superficial) bones develop directly from loose connective tissue. These are the incisive, maxillary, zygomatic, palatine, lacrimal, nasal, frontal, parietal, and interparietal bones of the skull.

2. The **endochondral** (cartilaginous or deep) bones of the skull are the vomer, presphenoid, basisphenoid, basioccipital, ethmoid, and ethmoturbinate. These are preformed in cartilage. The **occipital** and the **temporal bones** are each a combination of the two types of bone. The supraoccipital part of the occipital is membranous, and the occipital condyles are endochondral. In the temporal bone the process of the temporal, which forms the caudal half of the **zygomatic arch,** and the large, curved squamous portion are membranous, while the **petrosal** portion and the **bulla** are endochondral. Both the membranous and endochondral types have **haversian systems** when fully formed; they cannot be distinguished histologically when fully formed.

FORAMINA OF SKULL

1. The **foramen magnum** is the large opening in the occipital bone for the passage of the spinal cord to join the medulla oblongata of the brain. The spinal accessory nerves and the vertebral arteries also pass through this opening.

2. The **hypoglossal foramen** passes through the occipital bone at the inside lower portion of the foramen magnum, opens on the ventral surface with the jugular foramen, and transmits the **hypoglossal nerve.**

3. The **condylar foramen** opens on the inside upper part of the occipital condyle and transmits a vein from the **transverse sinus.**

4. The **jugular foramen** is at the junction of the tympanic bulla and the occipital bone in the caudomedial surface of the bulla. It transmits the **inferior cerebral vein** and the **ninth, tenth, and eleventh nerves.**

5. The **stylomastoid foramen** is between the stylomastoid process and the lateral caudal border of the bulla. It serves as a passage for a **branch of the seventh nerve.**

6. The **facial canal** passes through the medial part of the petrous portion of the temporal bone. It is the principal exit for branches of the **facial nerve.**

7. The **internal acoustic meatus** is below the facial canal on the medial surface of the petrous portion and serves as a passage for the **eighth (auditory) nerve** from the brain into the inner ear.

8. The **external acoustic meatus** is on the lateral surface of the tympanic bulla for the entrance of sound waves to the tympanum.

9. The **foramen for the eustachian tube** is lateral to the styloid process at the rostral edge of the tympanic bulla.

10. The **oval foramen** is at the basal caudolateral edge of the basisphenoid bone, medial to the mandibular fossa. It transmits the mandibular branch of the **trigeminal nerve** and **middle meningeal artery.**

11. The **round foramen** is in the base of the basisphenoid, rostral and slightly medial to the oval foramen, and serves as a passage for the **maxillary branch of the trigeminal nerve.**

12. The **orbital fissure,** or **foramen,** is between the presphenoid and basisphenoid bones laterally. Through this fissure pass the **third, fourth, and sixth cranial nerves** and the **ophthalmic branch of the fifth.**

13. The **optic foramen** is lateral in the pre-

sphenoid bone and transmits the **optic nerve.** Laterally and rostrally there may be a small **ethmoid foramen.**

14. The **olfactory foramina** pass through the cribriform plate of the ethmoid bone and permit the **olfactory nerves** to spread out on the turbinate bones of the nasal cavity.

15. The **sphenopalatine foramen** is the larger and more medial of the two in the palatine bone of the ventromedial wall of the orbit of the eye. It transmits the **sphenopalatine artery** and the **caudal superior nasal nerve,** which is a branch of the **fifth (trigeminal) nerve.**

16. The **palatine canal** is lateral to the sphenopalatine foramen and passes through to the rostrolateral edge of the palatine bone on the ventral surface of the hard palate. It transmits the **greater palatine nerve** and **artery.**

17. The **palatine fissure** is on each side of the median line, between the bases of the canine teeth, and is bounded by the incisive and and maxillary bones. The **nasopalatine nerve,** which is a branch of the maxillary, passes through it.

18. The **infraorbital foramen** is in the maxillary bone, below and rostral to the eye. It is large and transmits the **infraorbital nerve.** Sometimes two foramina are present instead of one, depending on where the nerve branches.

19. The **lacrimal canal** is between the lacrimal and maxillary bones in the rostromedial wall of orbit. The **lacrimal duct** passes through it on the way to the nasal chamber.

SOME DIFFERENCES IN SKULLS OF THE CAT AND MAN

1. Man has twenty-two separate skull bones, whereas the cat has thirty-five to forty.

2. Frontal and parietal bones are enlarged and pushed higher in man, whereas the jaws are relatively smaller and less protruding than those in the cat.

3. The incisive and maxilla on each side fuse into one bone in man but are separate in the cat.

4. The two frontal bones that are separate in embryo man become fused in the adult, but they remain separate in the adult cat.

5. The several parts of the sphenoid are fused into one bone in man but are in two principal parts in the adult cat.

6. The ossified dura mater forms a part of the parietal bone known as the tentorium in the cat but remains unossified in man.

7. A part of the hyoid branchial arch ossifies and forms the part of the temporal bone known as the styloid process in man but is not so formed in the cat.

8. The nuchal crest and the tympanic bullae are well formed in the cat but are absent in man.

9. An interparietal is often present as a separate bone in the cat but is only occasionally found in man.

10. The caudolateral wall of the orbit of the eye is well ossified in man but is only partially ossified in the cat.

11. Each half of the upper jaw of the cat has three incisors, one canine, three premolars, and one molar, whereas man has two, one, two, and three, respectively.

12. The inferior nasal conchae of man are separate bones, whereas in the cat these are parts of the maxillary bones, the maxilloturbinates or ventral nasal conchae.

13. The mandibles of the cat are easily separated from one another at the symphysis, whereas in man they are strongly fused.

14. The cat normally has thirteen pairs of ribs, but occasionally there are fourteen or even fifteen pairs. These extra ribs appear adjacent to the seventh cervical vertebra, adjacent to the first lumbar vertebra, or at both of these locations. In man there are usually twelve pairs of ribs, but extra pairs may appear at the same locations as those mentioned for the cat. Such an abnormal or unusual structure, which appears suddenly in an animal and is not present in its ancestors for several generations, is called an "anomaly." This anomaly, when it occurs by the first lumbar vertebra in man, is called a "gorilla" rib, since it is more often found in the gorilla.

REVIEW QUESTIONS ON THE SKULL

The student should answer all questions and hand them to the instructor for evaluation.

1. What are the two principal types or kinds of bones in the skull?

2. What are the names of two cartilaginous bones of the skull?

3. Name two endochondral bones of the skull.

4. Name the bones that help form the cranial cavity.

5. What is the general position of most of the foramina of the skull?

6. What bones help to form the orbit of the eye?

7. What is the explanation of the cleft palate?

8. Name and locate six foramina.

9. Within what bone are the parts of the inner and middle ear?

10. Locate and state the significance of the cribriform plate.

11. Name two air sinuses seen in the median sagittal section of the skull.

12. State five differences in the structure of the skulls of cat and man. Answer from your own observations.

13. About how many foramina are there in a cat's skull?

14. What structures usually pass through foramina?

15. What bones form the roof of the mouth?

16. What is an anomaly?

17. What bones constitute a zygomatic arch?

18. What structure helps to separate the cavities for the cerebrum and cerebellum?

19. What bones contain air sinuses?

20. What bones help to form the nasal passage?

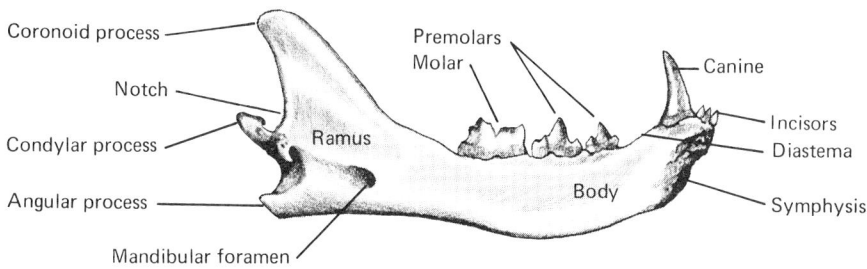

Fig. 1-7. Mandible, medial view.

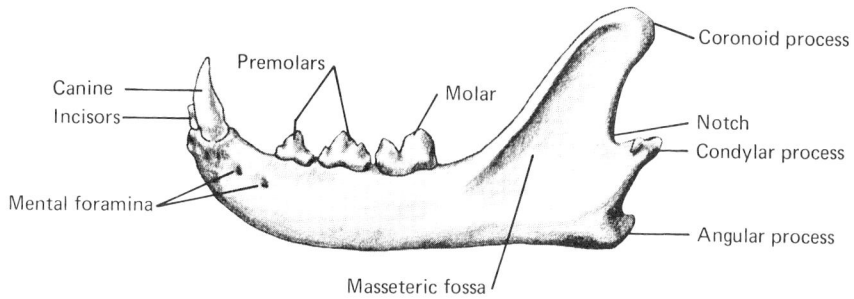

Fig. 1-8. Mandible, lateral view.

MANDIBLE (Figs. 1-7 and 1-8)

If you have two halves of the lower jaw, they were probably taken from different cats. Do you have the right or left jaw? Fit the jaw onto the skull. Examine the inner surface of one half of the lower jaw. The two halves are joined in the median line at their rostral extremities. This area of fusion is called the **symphysis.** Each half jaw consists of a horizontal portion, the **body,** bearing teeth on its alveolar border and an ascending portion, the **ramus.** In each half of the lower jaw there are three **incisors,** one **canine,** two **premolars,** and one **molar,** provided all are present. How do these teeth differ? The space between the canine and the first premolar is the **diastema.** Caudally the ramus is divided into a **condylar process,** which extends transversely and fits into the mandibular fossa to articulate with the skull, and a long, vertical **coronoid process,** to which the masseter muscle is attached. The lower proximal projection is the **angular process.** On the outer distal portion of the body are one or more **mental foramina,** which are the openings of the mandibular canal. This canal extends diagonally through each half jaw for the passage of the inferior alveolar artery and nerve. The latter is a subdivision of the mandibular branch of the fifth nerve. A small wire may be passed through some mandibular canals. The caudal end of the mandibular canal opens by a **mandibular foramen** on the inner surface of the ramus, near the base of the condylar process. The rostral portion of the canal continues to the incisors but also opens by one or more foramina lateral and ventral to the **diastema.** Above the condylar process is a **notch.** In man the two halves of the mandible are united at their cephalic ends to form a single bone. In the cat the two halves are separate but articulate closely at the **symphysis menti** by a thin interarticular cartilage. After the study of each cat bone, compare it with that of man, if the human bone is available.

Dental formula of the cat (*Felis*):

$$\text{i}\,\frac{3}{3},\ \text{c}\,\frac{1}{1},\ \text{pm}\,\frac{3}{2},\ \text{m}\,\frac{1}{1}\ \text{totals 30}$$

Dental formula of man (*Homo*):

$$\text{i}\,\frac{2}{2},\ \text{c}\,\frac{1}{1},\ \text{pm}\,\frac{2}{2},\ \text{m}\,\frac{3}{3}\ \text{totals 32}$$

Extra teeth (beyond the normal number) are seldom seen in the cat but are not so uncommon in man. They appear in the upper jaw, and sometimes there is almost a complete double set. This condition is called an "anomaly" and is interpreted as an **atavistic character,** a return to a remote ancestral condition.

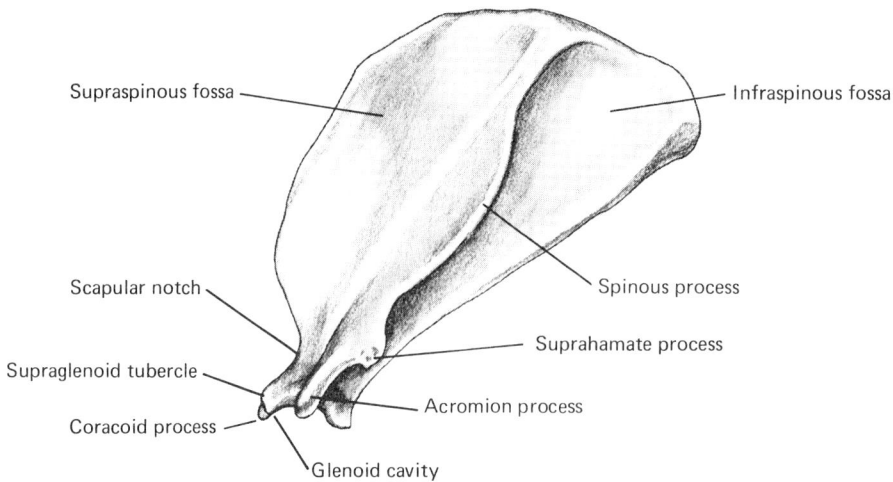

Fig. 1-9. Scapula, lateral view.

PECTORAL GIRDLE (Figs. 1-1 and 1-9)

The **scapula,** or shoulder blade, is held in place by muscles and articulates with the humerus. The **clavicle** is small and does not reach the scapula; in man it is relatively larger (see Fig. 1-2). On the lateral surface of the scapula is a long ridge, the **spinous process,** (**spine of the scapula**), which separates the **supraspinous fossa** from the **infraspinous fossa.** The spine has two projections at its lower extremity: the **acromion process,** which extends toward the **glenoid cavity** (socket for the humerus), and the **suprahamate process,** a flat, caudally directed process a short distance up on the spine. Near the upper inner edge of the glenoid cavity is the **coracoid process** of the scapula, which is short and curved. Immediately above the coracoid process on the cephalic edge is the **scapular notch.**

As you work on cat bones, compare them with those of man, if they are available. The **coracoid process** is vestigial in the cat and man but is a well-developed bone in most reptiles and in the bird, in which it is large and braces the shoulder. The shape of the scapula is well suited for the attachment of many muscles, which will be studied later.

HUMERUS (Fig. 1-10)

The proximal, or upper, end of the humerus is larger than the distal end and consists of a smooth, rounded central portion, the **caput** (**head**) and a lateral **greater tubercle,** which is separated from the head and **lesser tubercle**

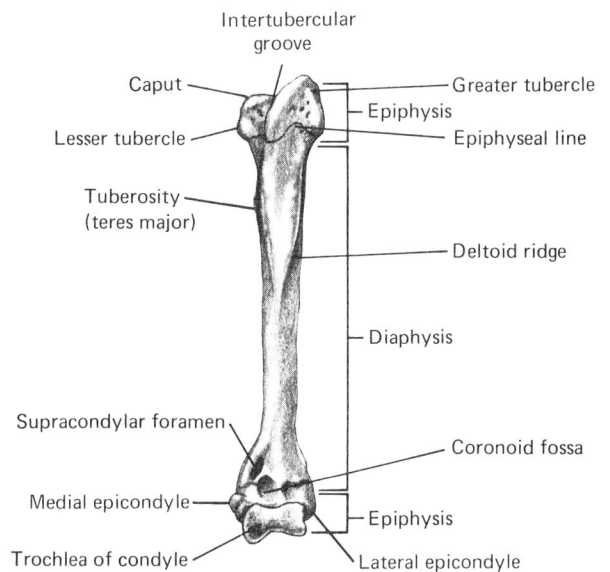

Fig. 1-10. Humerus, cranial view.

by the **intertubercular (bicipital) groove.** These parts constitute the **epiphysis,** the lower limit of which is the **epiphyseal line.** The **diaphysis** (**shaft**) forms most of the long medial smooth part, with the **supracondylar foramen** near the medial edge and distal extremity. On the cephalic surface of the shaft is a **crest,** which extends distally from the cranial limit of the **greater tubercle** down the cephalic surface. The **deltoid ridge** extends from the lateral limits of the greater tubercle distally and forward across the lateral surface to join the crest. On the medial surface, two-thirds of the distance from the proximal end, is a small **nutrient foramen**

14

for the entrance of a blood vessel. This is often difficult to locate. At the distal end on the posterior surface is a definite depression, the **olecranon fossa,** into which the **olecranon process** of the ulna (see Fig. 1-11), fits each time the forelimb is extended. The **distal articulating epiphysis** consists of the large, sharp-edged, circular **trochlea.** The **supracondylar arch** forms the inner boundary of the **supracondylar foramen,** which is absent in man. The **medial epicondyle** is distal and medial to the arch. Proximal to the **lateral epicondyle** is the **epicondylar crest.**

From the foregoing description, determine whether you have the right or the left humerus and compare it with the mounted cat's skeleton. Each long bone of the forelimb and hind limb consists of an elongate **diaphysis,** and at each end an enlarged articulating portion, the **epiphysis.** In early life the epiphyses are separated from the diaphysis by cartilage that continues to proliferate cells long after birth, which accounts for much of the increase in height during growth. Often, when the bones of a young animal are boiled to clean them, the epiphyses fall off, as may be noticed on some of the bones being studied. At the upper end of the fully developed humerus this area of growth is often easily seen and is known as the **epiphyseal line,** or **plate.**

RADIUS AND ULNA (Fig. 1-11)

The radius and ulna constitute the bones of the **forearm.** The smaller extremity of the radius at the proximal end is the **caput (head),** which contains a slight depression for articulation with the humerus. These bones articulate in a medium of synovial fluid. The depression has a rounded **tuberosity** at its lateral edge. The slight constriction distal to the head is the **collum (neck),** and immediately beyond it is the **tubercle,** which projects caudally toward the ulna. The **distal epiphysis** is larger than the proximal epiphysis and sometimes is easily separated from the **diaphysis,** which is slightly concave on its caudal surface. Laterally, an **articular facet** rubs against the ulna. The **distal epiphysis** possesses three grooves on its cranial surface for the passage of the extensor muscles of the forearm. The medial portion of the epiphysis also has a small, sharp **tubercle.**

The ulna has a large **semilunar notch** into which most of the distal end of the humerus

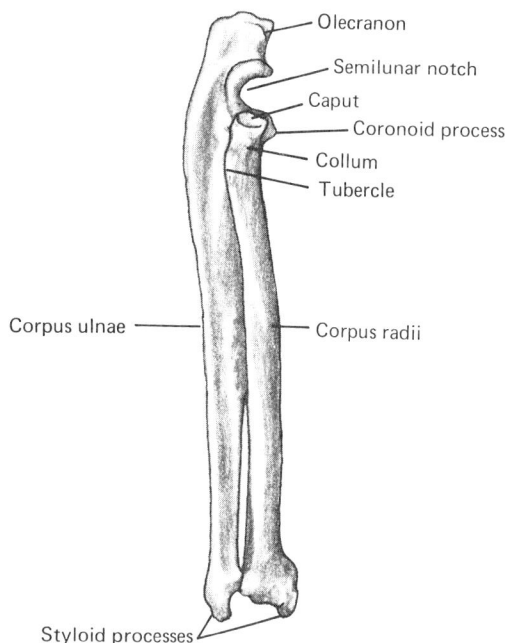

Fig. 1-11. Radius and ulna, medial view.

articulates. The medial edge of the notch is the smoother, since the lateral edge has a projection and a smaller notch for the head of the radius. The **coronoid process** projects distally from the lower ends of both notches. The large extention of the ulna at the elbow beyond the semilunar notch is the **olecranon process.** At the distal end of the ulna is the blunt-tipped **styloid process,** which has a slightly concave medial surface. The short projection near the base of the **styloid process** is the **articular facet,** for articulation with the radius. The styloid process represents the epiphysis, which if not well ossified may have come off when the bones were cleaned.

From the foregoing description, determine whether you have the right or the left ulna. If the humerus, radius, and ulna of your set are from the same side, fit them together. Examine and compare with mounted cat and human skeletons if available.

The bones of the thoracic and pelvic limbs of cat and man are considered **homologous,** since they are similar in origin and structure. They are **endochondral,** and the number of bones in comparable regions is almost the same. Sometimes there are extra toes on the forelimb of the cat, making a total of seven. The presence of extra toes is considered an **atavistic character.**

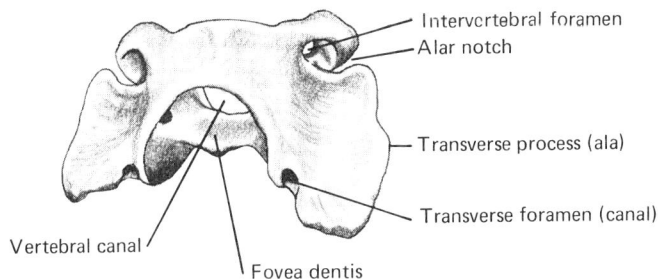

Fig. 1-12. Vertebra (atlas), caudodorsal view.

SPINAL COLUMN (CERVICAL, THORACIC, LUMBAR, SACRAL, AND COCCYGEAL VERTEBRAE)

The spinal column consists of a chain of somewhat similar bones called vertebrae. They are developed on the same general plan, but different degrees of ossification and variations in function have been important factors in causing them to appear differently in the highly specialized vertebrates, as in cat and man. There are five principal regions in the cat's spinal column. Representative vertebrae from each of these regions will now be considered.

CERVICAL VERTEBRAE

1. The **atlas,** or **first cervical vertebra** (Fig. 1-12), is distinctive in having large, flat **transverse processes,** a very small or no neural spine, and no centrum. It has a large **vertebral canal,** through which the spinal cord passes to join the medulla oblongata within the skull. The vertebral canal is bounded on the sides and above by the **neural arch** (Fig. 1-14), the dorsal portion of which may be determined by having more bony material. The **transverse processes** are winglike structures (**alae**) with small **transverse foramina** passing through the caudal part of each. The neural arch spreads at its cranial end to pass laterally to each occipital condyle of the skull, thus serving as **articular surfaces,** or **facets.**

Fit the atlas onto the skull in order to see the relationships. Immediately caudal and dorsal to each of the cranial articular surfaces is a small opening, the **cervical foramen,** for the passage of the first cervical nerve. Pass a small wire through each of the foramina of this vertebra in order to better determine the extent of each. The **posterior articular facets** are on the inner walls of the vertebral canal.

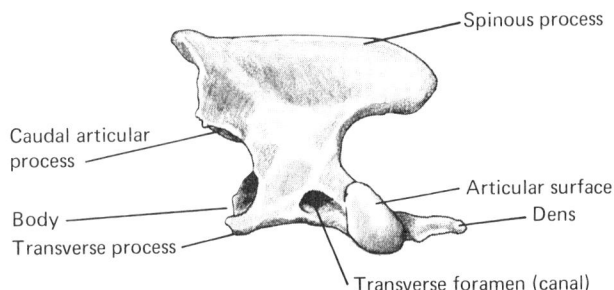

Fig. 1-13. Vertebra (axis), lateral view.

2. The **axis,** or **second cervical vertebra** (Fig. 1-13), is distinguished by having a long, narrow, cranial projection, the **dens** (**odontoid process**) and an enlarged, laterally compressed **spinous process.** The dens consists of most of the centrum of the first vertebra, which is fused with the centrum of the second, forming a toothlike projection. This process forms the axis around which the first vertebra turns when the head is rotated from side to side. The laterally compressed spinous process furnishes a large area for the attachment of muscles. A short **transverse process** surrounds a small **transverse foramen** on each side of the caudal portion of the **centrum.** The cranial **articular surfaces** (**facets**) are large, oval, smooth areas near the base of the dens. For the most part the facets face laterally and only slightly upward. The **caudal articular processes,** or zygapophyses, have facets that face downward. The cranial and caudal ends of a vertebra may be determined by the way these facets face; however, articular facets do not develop on all vertebrae. Fit the atlas onto the axis and the atlas onto the skull; observe that when the skull is rotated, there is a movement between the atlas and the axis. Movement between the skull and the atlas is of the hinge type.

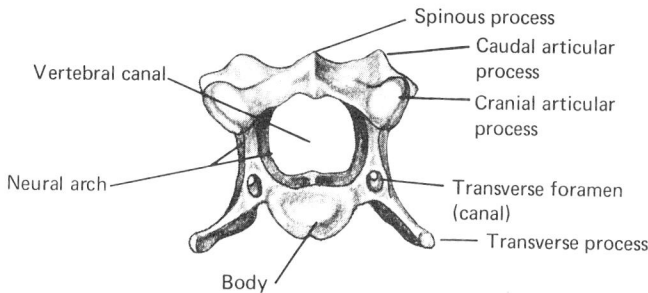

Fig. 1-14. Typical cervical vertebra, cranial view.

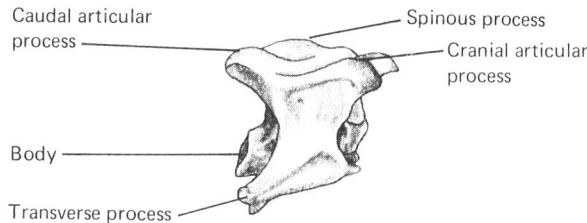

Fig. 1-15. Typical cervical vertebra, lateral view.

If possible, observe these structures on the human skeleton.

3. A **typical cervical vertebra** (Figs. 1-14 and 1-15) is represented by the third, fourth, fifth, or sixth vertebra. There are **seven cervical vertebrae** in mammals. Thus, a cat, elephant, giraffe, and man have the same number. The most distinctive structural characteristic that is found on the majority of cervical vertebrae is the **transverse foramen** within the base of the transverse process on each side of the **body** (**centrum**); however, the seventh cervical vertebra of the cat does not have this foramen because the **vestigial rib** that completes the foramen is not fused with the transverse processes. In other words there is a **remnant of a rib** on each of the first six cervical vertebrae in all mammals that completes the **transverse foramen** and projects as an extension of the **transverse process**. The **neural arch** surrounds the **vertebral canal** dorsal to the centrum, and projecting from its dorsal surface is the **spinous process** (**neural spine**), which varies in size and length in different vertebrae. The **neural arch** surrounds the **spinal cord**. The centra of adjacent vertebrae are typically separated from one another by an **intervertebral disk**. In addition to this disk there are usually specialized areas of contact where parts rub against one another. These areas of contact are called **articular processes** and **articular facets**.

It is necessary to be able to determine the cranial and caudal extremities of vertebrae. This can be done by studying the **articular facets**. Examine a typical cervical vertebra. Hold it with the spinous process upward. The **cranial articular** facets that you see **face upward and slightly toward the midline**. The two projections that bear these facets are the **cranial articular processes** (**prezygapophyses**). On the opposite extremity are the **caudal articular processes** (**postzygapophyses**), with facets that **face downward and slightly laterally**. Fit adjacent vertebrae together or examine a mounted skeleton and see how the facets articulate with one another. The **cranial articular** facets are usually farther apart than are the caudal ones.

THORACIC VERTEBRAE AND RIBS
(Figs. 1-1 and 1-16)

The general structural characteristics of a typical thoracic vertebra include articular facets (as in the cervical), facets for the ribs, and **spinous processes** (**long neural spines**), which usually **project dorsocaudally**. This slope gives the spines a better brace against the pull of the muscles that hold up the head. The neural spines of the last four thoracic vertebrae project slightly forward, since their principal muscles extend back to the pelvic region and the pull on the spines is in the opposite direction. This is better seen on the mounted skeleton of the cat. These

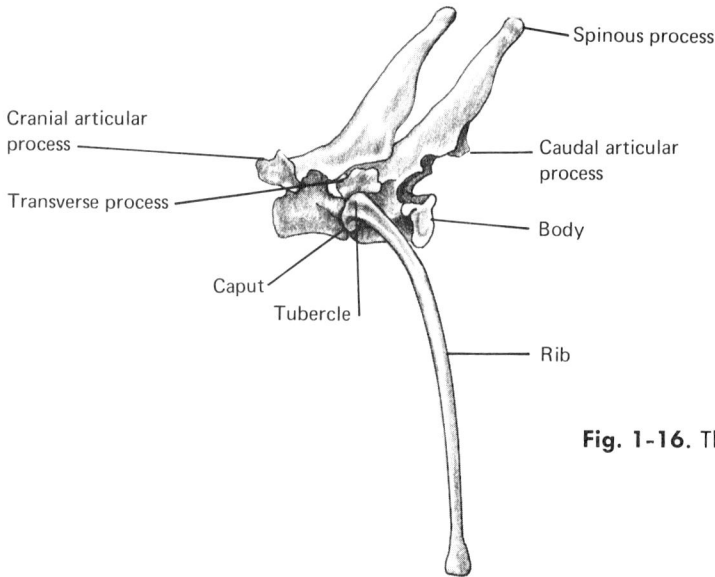

Fig. 1-16. Thoracic vertebra and rib, lateral view.

spines project similarly in man, even though man carries his body in an erect position. This suggests that his ancestors may have carried their bodies in a horizontal position. The prezygapophyses and postzygapophyses with facets are present in the thoracic vertebrae, as they are in the cervical vertebrae. Each thoracic vertebra has two short articular transverse processes on each side. The upper, or **diapophysis,** projects laterally from the neural arch, whereas the lower one, the base of the **parapophysis,** is reduced to an articular surface on the side of the centrum. In fact, parts of two articular surfaces may often be seen on one side of the centrum, because the head of the rib joins the vertebra in such a manner as to touch two vertebrae. Each half of an articular surface on the centrum is a **demifacet.** Compare a thoracic vertebra with a cervical vertebra, which was examined previously, for homologous structures.

The proximal end of the **thoracic rib** bears a close relationship to the lateral articular surface of the thoracic vertebrae just mentioned. At the extreme upper end of the rib is an articular surface known as the **caput (head),** which joins with the **parapophysis** of the centrum. A small projection on the rib close to the caput is the **tubercle,** and this likewise joins, or articulates with, the **diapophysis** of the vertebra. The **collum (neck)** of the rib is the constriction between the caput and the tubercle.

Fit the caput and tubercle of the rib against the parapophysis and diapophysis, respectively.

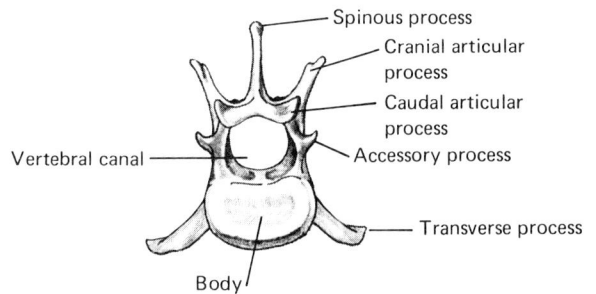

Fig. 1-17. Lumbar vertebra, caudal view.

The space formed between these four parts is homologous with the transverse foramen of a cervical vertebra. Hence, the part that completes the transverse foramen in the neck vertebrae is really the proximal end of a rib. On the lateral surface of a typical rib is a projection called the **angle of rib.** This is not present on all the ribs of all cats. The rib continues ventrocaudally, and the **shaft** joins what is usually a rather long costal cartilage that is intermediate between the rib proper and the sternum. In different cats this cartilage may be found in various degrees of ossification. It forms an angle with the distal end of the rib.

LUMBAR VERTEBRAE (Fig. 1-17)

The lumbar vertebrae are distinctive in having large, long **transverse processes** that project in a craniolateral direction. This is probably the result of their response to the pull of the more important muscles attached to them. These

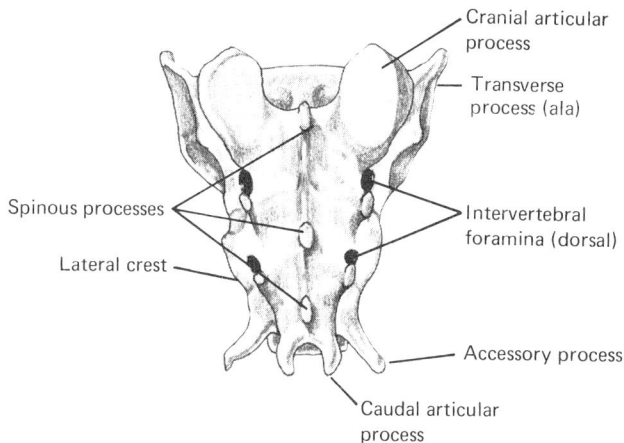

Fig. 1-18. Sacrum, dorsal view.

Labels on figure:
- Cranial articular process
- Transverse process (ala)
- Intervertebral foramina (dorsal)
- Accessory process
- Caudal articular process
- Spinous processes
- Lateral crest

muscles extend to the pelvic girdle and to the pelvic limbs. The smooth, cup-shaped articular surfaces, or facets, of the **cranial articular process** (**prezygapophysis**) face toward the center and upward, whereas the articular surfaces of the **caudal articular process** (**postzygapophysis**) face outward and downward. Below these on each side is often an **accessory process** (**anapophysis**) that helps to lock these vertebrae more firmly together and thus makes the back stronger. The lumbar vertebrae, and especially their **centra**, are relatively large. The heavy but short **neural spine** projects slightly forward, and the **vertebral canal** is large.

The first lumbar vertebra of man sometimes has a **vestigial rib**. This rib is normal in the gorilla, hence it is known as the "gorilla rib" when it occurs in man. It is considered an "anomaly."

SACRAL VERTEBRAE (Fig. 1-18)

The **sacrum,** or synsacrum, consists of three vertebrae fused into one bone; however, the outline of each vertebra is easily seen on the ventral surface. Two pairs of **dorsal intervertebral foramina** and two pairs of **ventral intervertebral foramina** are present, the size of each depending on the extent of ossification or fusion of transverse processes. These foramina represent the spaces between transverse processes of vertebrae and serve for the passage of branches of sacral nerves. The cranial processes are the **prezygapophyses** with small, smooth **articular facets**, and the transverse processes are large, winglike projections (**alae**), which because of the evidence shown in their embryological development, are believed to be **vestiges** of ribs that

articulate with the ilium. At the caudal extremity of the sacrum, the projections near the vertebral canal are the **postzygapophyses** with articular facets. Three **neural spines** extend dorsally, and lateral to these spines on each side there is a row of four **tubercles** that converge caudally. These are **fused prearticular** and **postarticular processes** (**zygapophyses**). The cartilage that separates the **lateral crest** (**pars lateralis and wing**) from each **ilium** relaxes in the female when she gives birth to the young. According to H. H. Wilder,* "the sacral region of man consists normally of the fusion of five vertebrae, but there are records of seven, which should serve to dispel the idea that the body of man (or any other animal) is formed in accordance with a definite pattern, or is constructed on any other principle, save those of heredity and environment."

COCCYGEAL VERTEBRAE
(Figs. 1-19 and 1-20)

The coccygeal, or caudal, vertebrae are very different in structure from the base to the tip of the tail. This variation depends on the extent of ossification of the various vertebrae. Fig. 1-19 shows a vertebra close to the base of the tail.

The **first ten vertebrae** in this region are **most nearly typical** for the entire spinal column, since most of the major parts found in any vertebra are present. The **neural arch** and the **cranial** and **caudal zygapophyses,** with their articulating surfaces and short neural spines, are well formed. The **transverse processes** are unusually well developed for the size of the vertebrae, and on

---

*Wilder, H. H.: History of the human body, New York, 1923, Henry Holt & Co., p. 131.

19

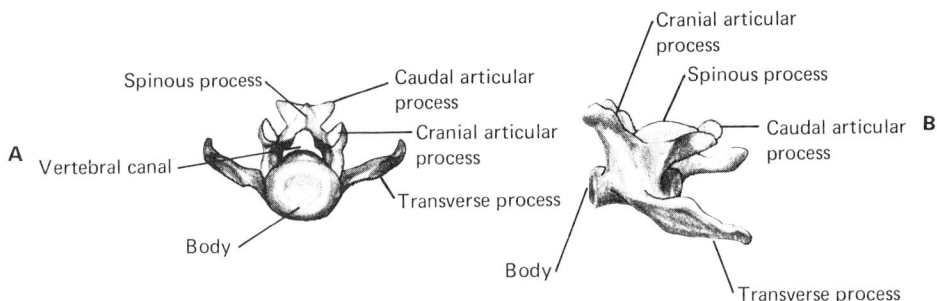

Fig. 1-19. A, Cranial coccygeal vertebra, cranial view. **B**, Cranial coccygeal vertebra, lateral view.

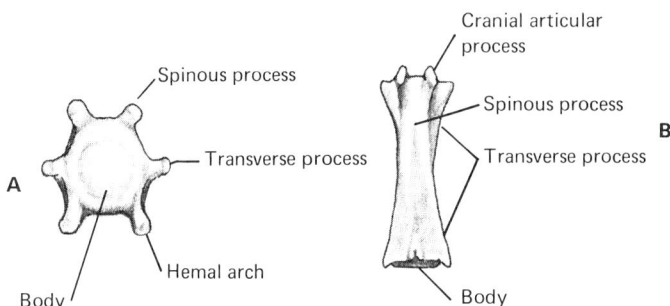

Fig. 1-20. A, Caudal coccygeal vertebra, cranial view. **B**, Caudal coccygeal vertebra, dorsal view.

the ventrocranial ends of the centra are what are interpreted as remnants of the **hemal arches**, known as **chevron bones.** There may be several of these, beginning with the **fifth vertebra.** No other vertebrae of the cat have hemal arch remnants, and it is especially because of these remnants, in addition to the other structures of the vertebrae, that the first eight or ten caudal vertebrae are considered **most nearly typical** of all of the vertebrae of the cat's spinal column.

Fig. 1-20 shows a vertebra of the caudal portion of the tail, where the **centrum (body)** is almost all that remains of the vertebra; however, remnants of the bases of the **neural** and **hemal arches** are present. The bases of the former are farther apart and longer. Parts of the **transverse processes** are also present. When eating canned salmon, you often find partially calcified vertebrae in which the centra is about all that is left. In man there are remnants of only three to five centra remaining of the caudal vertebrae.

When you look at the spinal column of the cat from the side, as in Fig. 1-1, you observe a long arch, but when you look at the spinal column of man in an erect position and from the right side, you see the spinal column in a form somewhat like the capital letter S, which serves like a spring in supporting the body. As man becomes older, this shape tends to change and becomes somewhat like the capital letter C, as in a stooped old person. Examine the mounted skeletons of cat and man.

SOME DIFFERENCES IN VERTEBRAL, OR SPINAL, COLUMNS OF THE CAT AND MAN

1. The number of vertebrae:

Regions	Cat	Man
Cervical	7	7
Thoracic	13	12
Lumbar	7	5
Sacral	3	5
Coccygeal	4-26	3-5

2. Thoracic vertebrae in the cat have relatively longer and larger neural spines or processes than in man.

3. Transverse processes on lumbar vertebrae are relatively longer and project toward the skull at a more acute angle in the cat than in man.

4. Vestigial chevron bones occasionally project ventrally from the cranial ends of the fifth

to the thirteenth coccygeal vertebrae in the cat. These are absent in man.

5. The first eight or ten coccygeal vertebrae are much more typical than any others of the cat. The few coccygeal vertebrae present in man are vestigial and are mere remnants of typical vertebrae.

6. Man has twelve pairs of ribs; the cat has thirteen.

7. Man has seven pairs of true ribs; the cat has nine. True ribs are attached separately to the sternum.

8. Man has five pairs of false ribs; the cat has four. False ribs are not attached directly to the sternum.

9. Man has two pairs of floating ribs; the cat has one pair. Floating ribs are those that are not attached at their ventral ends.

10. In man the body of the sternum is fused into one piece, but there are four sternebrae in the human embryo. In the cat there are eight sternebrae in a continuous line.

11. The suprahamate process of the cat scapula is absent, as such, in man.

12. Normally five vertebrae fuse to form the sacrum of man; three vertebrae fuse in the cat.

13. In the cat there are seven lumbar vertebrae but only five in man.

Name _____

Date _____

REVIEW QUESTIONS ON SPINAL COLUMN

1. What are the parts of a typical cervical vertebra?

2. Which of all the vertebrae of the cat are most nearly typical?

3. What are the principal distinctive, structural characteristics of the vertebrae in each of the five regions in the spinal column of a cat?

4. What are the structural differences between the first two vertebrae of the neck and the other cervical vertebrae?

5. What are some of the factors that probably help to determine the direction in which neural spines and transverse processes project?

6. How do the vertebrae of the sacrum differ in a cat and man? How many vertebrae are a part of the sacrum in the cat? How many in man? Does the erect position demand more vertebrae?

7. Which vertebrae of the cat have the longest neural spines or processes? Why?

8. What general statements may be made concerning the differences in the coccygeal, or caudal, vertebrae of the cat?

9. What parts of the spinal column or vertebrae do we often find when eating canned salmon?

10. How may the cranial or the caudal end of a vertebra be determined by studying the articular surfaces, or facets?

11. How does the number of vertebrae differ in the cat and in man?

12. Which vertebra has no centrum? (Observe mounted skeletons.)

13. Would you say that the vertebrae of the cat and man, in comparable regions, are quite similar in structure? (Observe mounted skeletons.)

14. Are all vertebrae of cat and man built on the same general plan? (Compare mounted skeletons of each.)

15. Are the thoracic or the pelvic limbs and girdles of the cat more firmly attached (normally) to the spinal column? (Observe the mounted skeleton.)

16. Name the parts that surround the transverse foramen of a typical cervical vertebra.

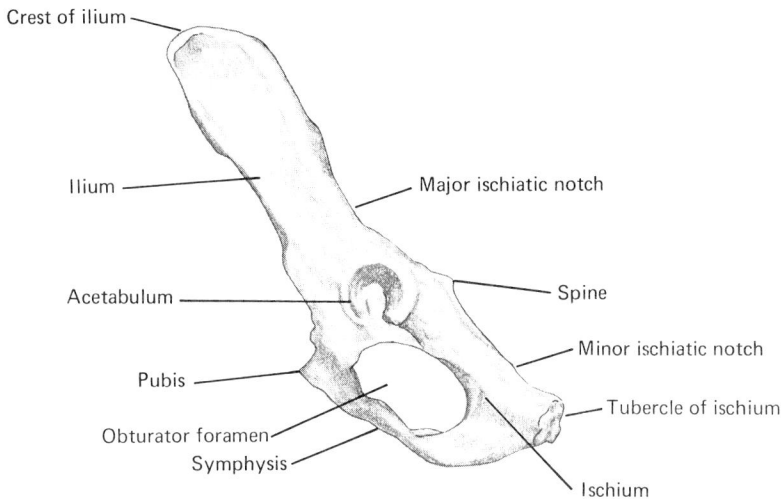

Fig. 1-21. Pelvic girdle, lateral view.

PELVIC GIRDLE (Fig. 1-21)

Each half of the pelvic girdle is called the **os coxae** and consists embryonically of three separate parts: ilium, ischium, and pubis. In a young mammal the lines of union of these three parts may be seen easily. The **crest of the ilium** is at is cranial extremity, and the **tubercle of the ischium** is at its caudal end. The **acetabulum** is the cup for the articulation of the femur. The large space separating the pubis and ischium is the **obturator foramen.** On the dorsal surface, close to the acetabulum, is the **spine of the ischium.** Cranial to this spine is the **major isciatic notch,** over which passes the isciatic nerve. Caudal to this spine is the **minor isciatic notch,** over which passes the tendon of the obturatorius internus. The right and left portions form a symphysis on the midline.

The ligamentous attachment between the sacrum and ilium firmly connects the hind limbs with the spinal column; this enables quick locomotion of the whole body, particularly in quadrupeds. The **pectoral girdle,** however, is attached to the body or its bones only by muscles. The muscles, as well as the bend of the limb at the joints, act as a spring or shock absorber when the cat lands on the front limbs following a jump.

FEMUR (Fig. 1-22)

The proximal end of the femur has a pronounced, rounded **caput (head)** that fits into the acetabulum of the pelvic girdle. Near the

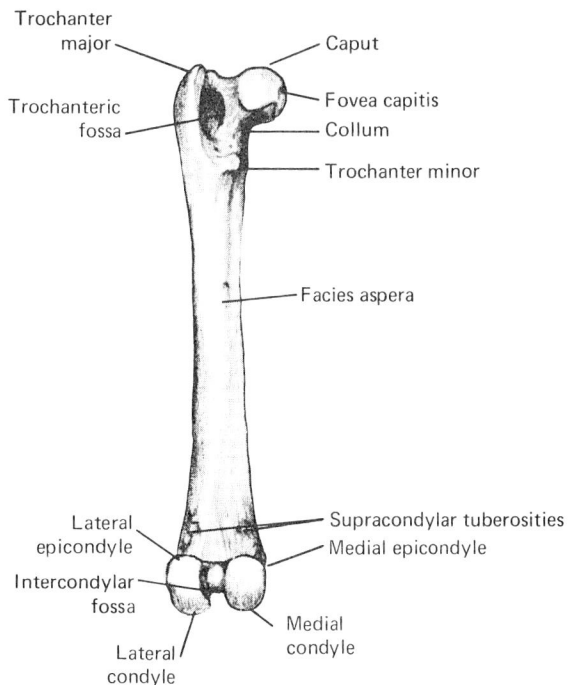

Fig. 1-22. Os femoris, caudal view.

center of the caput is a depression, the **fovea capitis.** The constriction adjacent to the head is the **collum (neck).** The larger tuberosity at the proximal end is the **trochanter major;** the pronounced depression on the caudal surface is the **trochanteric fossa.** The **trochanter minor** is immediately below this fossa. The rough line distal to the trochanter minor and extending diagonally is the **facies (linea) aspera.** The

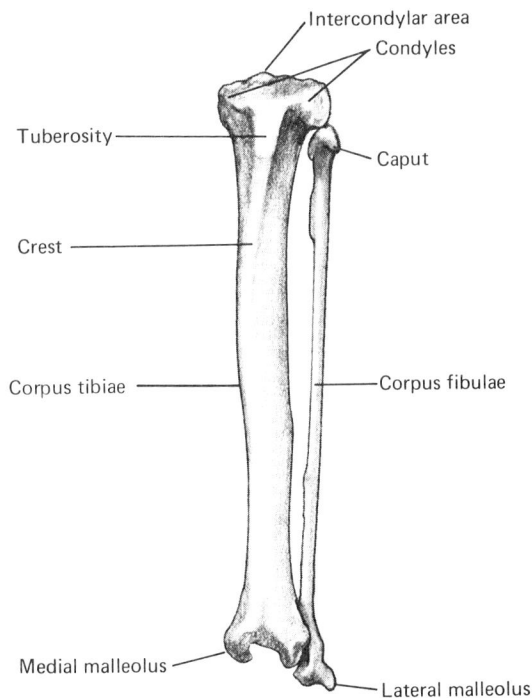

Fig. 1-23. Tibia and fibula, cranial view.

Labels on figure: Intercondylar area, Condyles, Tuberosity, Caput, Crest, Corpus tibiae, Corpus fibulae, Medial malleolus, Lateral malleolus

distal extremity is marked by the **lateral condyle** and the **medial condyle**, which are separated by **intercondylar fossa.** There is a small **lateral epicondyle** adjacent to the lateral condyle. The right and left femurs may be determined by the fact that the head of the femur is on the medial side and the posterior surface of the femur is slightly concave; in addition the condyles at the distal end project caudally.

TIBIA AND FIBULA (Fig. 1-23)

At the proximal end of the **tibia** the **epiphysis** is quite large and is easily seen. It separates off easily when bones of a young cat are boiled in cleaning. It consists of a **lateral condyle** on the more concave side and has a **facet** for the articulation of the fibula. The **medial condyle** is opposite the external condyle and is slightly higher. Between the two articular surfaces for the femur is an **intercondylar projection.** A small, blunt **tuberosity** is on the cranial edge below the epiphysis for the attachment of the patellar ligament. Below this is the **crest** of the tibia (shinbone), which extends downward and gradually decreases as the shaft becomes more nearly cylindrical. The distal end has an irregular depressed articulating surface, the inner wall of which forms a projection, the **medial malleolus.** A lateral projection has a **facet** at its base for articulation with the fibula.

From the foregoing description determine whether you have a right or a left tibia.

The **fibula** is well ossified only on older specimens; hence the **epiphyseal** ends come off easily on young cats. You should have a well-ossified fibula in order to identify the following parts: a **facet,** on the medial surface of the proximal end, for articulation with the lateral condyle of the tibia; the **diaphysis (shaft),** a long, irregular, and slender bone; a **pronounced groove** for the passage of the tendon of the peroneus longus muscle at the distal end; and the **lateral malleolus,** a projection on the outer boundary of the groove. On the medial surface of the distal end of the fibula is a large articular surface for the calcaneus, or heel bone, and above it is a smaller triangular facet for the tibia.

From the foregoing description determine whether you have the right or the left fibula. Fit the femur, tibia, and fibula together properly. This can be done only if they all belong to the right side or all to the left side. See a mounted skeleton.

GROWTH OF LONG BONES

Each of the three **endochondral** long bones of each leg has an **epiphysis** at each end, separated from a longer shaft, or **diaphysis,** by a **cartilaginous area** (see Fig. 1-10). Because of this cartilage, the epiphyses, or ends, of the long limb bones of a young animal come off easily. It is on the **shaft side of this cartilaginous area** where the **greatest cell division** or **growth** takes place. This growth adds to the length of the **diaphysis** by **pushing the epiphysis away,** thus adding to the length of the entire bone. This is how the limbs elongate and the height of the animal increases. Cartilage **does not** turn into bone but is absorbed and replaced by osteogenic cells brought in by blood vessels, which produce the bone. This, quite briefly, is how **endochondral,** or **cartilaginous, bones** grow. The **membranous,** or **dermal, bones** of the skull form directly from connective tissue and do not pass through a cartilaginous stage.

SOME DIFFERENCES IN SKELETONS OF THE CAT AND MAN

As you read the following summary, refer to mounted skeletons of cat and man, if available, in order to better see their similarities and differences.

1. Man has 126 bones, whereas the cat usually has 116 to 148. There is much variation in the tail of the cat.

2. The clavicle of the cat is relatively smaller and is not in contact with the manubrium or the scapula, as in man, but is embedded in muscle.

3. A supracondylar foramen is present on the humerus of the cat but is absent in man.

4. The pubic symphysis is better ossified in the cat than it is in man.

5. The cat has seven carpal bones, whereas man has eight.

6. Man has five fingers and five toes normally. Occasionally there is a reduction in the number of fingers. However, there are records of the human baby having six toes on each foot. The cat has five in front and only four on each hind foot. Occasionally the cat has six toes on each front foot.

7. Claws are on the ends of the toes of the cat, whereas nails are on the ends of the toes of man. However, claws are retractable.

8. The cat walks with only the digits touching the ground and therefore is a digitigrade, whereas man walks with his digits, sole of the foot, and heel touching the ground and therefore is a plantigrade.

9. The cat's spinal column is carried in an arched or a nearly horizontal position, whereas that of man is almost vertical. The nearly horizontal position, as in the cat, is better for the most efficient functioning of the internal organs, particularly the digestive, urogenital, and circulatory systems.

10. The vertebrae of the cat are much more varied in structure in different regions than those in man. The long neural processes of the thoracic region are for the attachment of mus-

cles that support the head on a nearly horizontal neck. The caudal vertebrae of the cat are much more numerous, whereas in man they are reduced to three or four.

11. The shapes of the skulls in man and cat are quite different, largely because of the great enlargement of the cerebrum in man that pushed the frontal and parietal bones dorsally and the nose ventrally or anteriorly.

12. On the ventral surface of a few of the caudal vertebrae of the cat are chevron bones, interpreted as being remnants of hemal arches.

13. The transverse processes of the lumbar vertebrae of the cat project sharply forward for the attachment of muscles extending caudally to the pelvic girdle. This condition is more pronounced in animals that climb trees.

14. In the cat, the length of the femur is correlated with the space between the acetabulum and the last rib, so that when a cat jumps, the distal end of the femur will not hit the rib, Man does not bring the knee as close to the ribs, and the space is not equal to the length of the femur.

15. A kitten is able to walk in a few days after birth, a calf in a few hours, but a human child is usually about one year old before he is able to walk. It is much easier to balance on four legs rather than two. Compare Figs. 1-1 and 1-2, particularly the way the hind feet come in contact with the ground. The same mechanical principle is involved in a four-wheeled wagon and a two-wheeled cart. The flat foot of man helps to compensate for the lack of four feet. The erect position of man requires a stronger sacrum and is much more unstable and difficult to maintain on two legs than the horizontal position of the body with four legs for support. When man first learns to locomote, he goes on his hands and knees, which is equivalent to four legs.

REVIEW QUESTIONS ON PELVIC AND LIMB BONES

1. What are the basic or fundamental parts of the pectoral girdle of the cat or man?

2. Which of the cat's girdles is more firmly attached to the spinal column, and what is the advantage of each type?

3. What are the advantages of having two limbs or four limbs?

4. State some of the advantages and some of the disadvantages of the erect position of the body of man.

5. Which performs the greater function, the clavicle of a cat or of man? (To answer, use your own judgment. Consider the movements of the forelimbs.)

6. Are limb bones endochondral or membranous?

7. Name ten structural differences in the skeleton of a cat and man.

8. What thoracic and pelvic limb bones in the cat and man are homologous? What facts indicate that they are so?

9. Which bones of the skull are membranous and which are endochondral in origin?

10. What are the remnants of the hemal arches called and where are they found?

11. What bones merge to form the os coxae (pelvis)?

12. How are the claws of a cat arranged so that they may be used effectively or not used on an instant's notice? (Examine a live cat.)

13. What is the condition and apparent function of the coracoid process in the bird, cat, and man?

14. Give an example for each of three functions of bones.

15. Name two bones from different groups that function in all three ways.

16. Briefly, how do the long bones of the limbs increase in length?

17. Does the cat or man walk and run more nearly on his toes? Explain. (Rely on your own observation for the answer.)

TWO
Muscles of the cat

Dissection of the cat is one of the best preparations for dissection of the human body, which is required in the first year of all medical and dental schools. Dissection of the cat is strongly recommended for nurses and physical education students, because after careful dissection one remembers by visualization, which is the best-known way to learn anatomy. In doing this work it is best for the student to read, previous to each laboratory period, the information that will be covered.

Most cats prepared for dissection have not had the skin removed. This may be done in different ways, but I prefer the following method. Cut through the skin down the center of the back, from the base of the tail to the top of the head, and dissect it from the left side by cutting the loose connective tissue, or fascia. On the side of the head and neck the fine fibers of the **platysma muscle** will be cut and some will come off with the skin, but most fibers remain on the neck. Make a transverse cut of the skin from the top of the scapulae, down on the lateral surface of the left thoracic limb. Cut the skin around the ear and eye and to the corner of the mouth. As the skin is being removed from the side of the arm and shoulder, the **cephalic vein** should be seen. Make another cut of the skin down the side of the left pelvic limb. Dissect the skin from the left side of the body to the midventral line, or **raphe.**

If the cat was pregnant or nursing kittens, the **mammary glands** can be seen along the ventral surface between the skin and underlying muscles. There are usually four or five pairs of nipples and mammary glands. The cranial mammary glands are supplied by the **mammary artery,** which in the female extends below the diaphragm and is called the **cranial epigastric**
artery. The caudal mammary glands are supplied by the caudal **epigastric artery.**

In the elephant, bat, monkey, and man the mammary glands are usually confined to the **pectoral region,** but in the horse, cow, sheep, and goat these glands are in the **inguinal region.** In the dog, cat, pig, and rat, they are all along the ventral surface. Occasionally in man there are cases in which five or more irregular pairs of nipples occur.

If your cat is a male, carefully dissect the skin from around the testes, penis, and sperm ducts, leaving them uninjured. If the cat has been properly injected and embalmed, there is little danger that it will become unfit for dissecting because of decomposition. If there are signs that decomposition has begun, more embalming solution can be applied to the surface. If the specimen is to be worked on over a period of several weeks, there is some danger of it drying too much. To prevent drying, the skin, which is usually still attached at several areas, may be pulled over the specimen between periods of study, or the specimen may be wrapped in wet newspaper and tied up securely or placed in a plastic bag. If a plastic bag is used, it is best to clip the claws.

Myology is the science that treats of the form, location, and attachments of muscles. The end of the muscle nearer the median plane of the body or farther up on the limb is known as its **origin,** whereas the place of attachment to the bone that it moves is known as its **insertion.** Usually skeletal muscles arise and terminate in white fibrous tissues known as tendons. The skeletal muscles are usually named according to their function, shape, or parts to which they are attached. Muscles are covered by **fascia** called **epimysium,** which also binds them together and which must be broken in order to separate the muscles. **Deep fascia** lies close against the muscles and dips down between them. **Loose fascia** lies under the skin and holds it to the muscles. Tendons, ligaments, and aponeuroses are forms of **heavy fascia,** and each is a form of connective tissue.

The **aponeuroses** and flat, white tendons are the stronger portions of this fascia. A **raphe** consists of the ridge or furrow along the line uniting halves of symmetrical parts, as down the middle of the back or of the chest or abdomen. The **raphe** indicates the position of the median plane of bilateral symmetry.

Dissection consists of **intelligent separation** of one structure from another and the **reflection** of known parts in order to study those lying deeper in the specimen. When reflecting a muscle, cut it into two parts transversely, turn each part back to expose deeper organs, but always retain the **origin** and **insertion** of each muscle studied. **Bisect** means to cut transversely into two parts. This cut is usually made near the center or middle of the muscle.

The **skeletal muscles** arise from myotomes (myomeres) of the mesoblastic, or mesodermal, embryonic sections of mesoderm (somites). Muscles are combinations of these myotomes, which are seen in various stages of fusion. Few muscles arise from a single somite. Basically, a pair of spinal nerves supplies each pair of somites. The **intercostal muscles** represent individual somites, but these are split horizontally into the external and the internal. Most muscles in the cat and in man are formed by merging, or fusion, of several somites, and each, with few exceptions, retains its **original nerve;** however, these nerves merge with one another and may again separate, which complicates identification.

Histologically, there are three kinds of **muscles.**

1. **Skeletal** muscle, which is **striated,** constitutes the flesh of an animal and is under the control of the will. Skeletal muscles are the most numerous and produce most of the movements. There are four different groups of skeletal muscles, depending on their location, as discussed in the following section.

2. **Cardiac muscle,** which is **striated,** constitutes the heart and is involuntary.

3. **Smooth muscle,** which is **nonstriated,** is located in the walls of the blood vessels and digestive tract.

FOUR MAJOR GROUPS OF SKELETAL MUSCLES

1. **Integumentary,** or **cutaneous, muscles** are attached to the underside of the skin and are mostly removed with the skin.

2. **Appendicular muscles** constitute the meat, or flesh, on the girdle and limb bones and are perhaps the most outstanding of the four major groups.

3. **Axial muscles** lie along the spinal column and are attached largely to the bones or vertebrae of the spinal column.

4. **Branchial muscles** were originally between the cartilages or bones of the gill arches and were used to move the gills and hence, indirectly, were used in respiration; in cat and man they lie between the bones and cartilages that constitute the lower jaws, larynx, and manubrium.

The following groups of muscles are arranged for convenience in dissection and study. The majority of them are appendicular in that they are associated with the girdles and limbs, but the other three major groups are also represented.

The description of each muscle is general, and the student is left to work out the details. There is a considerable variation in muscles, so do not expect the descriptions to be accurate in all details. Begin the dissection of the muscles on the left side of the cat. Turn the specimen onto its right side with its head to your left and the feet directed toward you.

INTEGUMENTARY, OR CUTANEOUS, MUSCLES

Most skin muscles are removed with the skin, but parts remain on your specimen. Identify these remnants on the left side of the cat and loosen their edges. They are very thin and appear as fine lines.

1. The **cutaneus trunci** (Fig. 2-1) is thin and covers most of the side of the body. It lies between the skin and the muscles of the body wall and is usually dissected off with the skin. Identify this muscle and determine the direction in which its fibers extend in different regions. It becomes thicker in the lateral wall of the chest where it fuses with the **latissimus dorsi** in the axillary region. If the skin has been removed, portions of the cutaneus trunci may be found closely attached to the latissimus dorsi ventral to the scapula. This extremity is considered as its origin, and the insertion is on the underside of a large area of

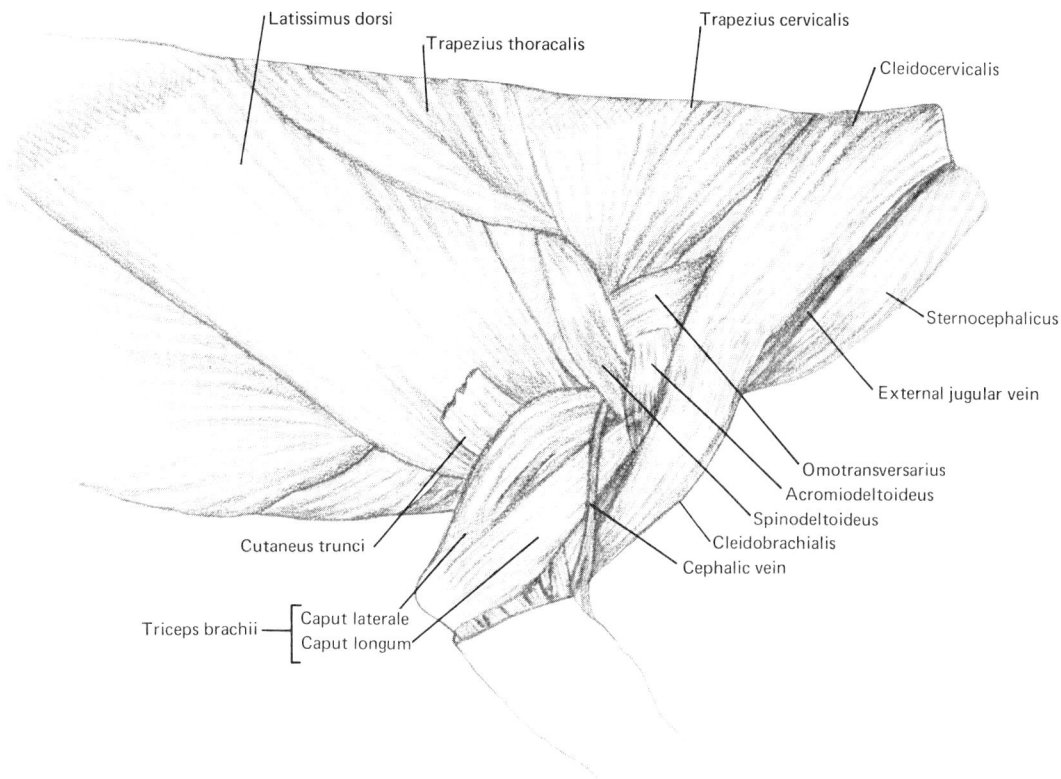

Fig. 2-1. Superficial muscles of neck, shoulder, arm, and thorax.

the skin. The cutaneus trunci causes much of the hair to stand on end when the cat becomes frightened; it is not found in man.

2. The **platysma** lies on the lateral side of the neck and head; some fibers may extend to the angle of the mouth. Many of these fibers are removed with the skin, but some may be seen covering the cranial end of the large, external jugular vein of the neck. This muscle moves the skin of the lateral region of the neck and face; it is much the same in man.

SUPERFICIAL MUSCLES OF NECK AND SHOULDER (Fig. 2-1)

Place the cat before you with its head to your left and its feet toward you.

Get the left scapula and humerus from your set of bones and check the names of their parts that are labeled in the drawing before you begin your study of the muscles. Place these two bones in their relative positions on your cat. Look at the clavicle on the mounted skeleton; this will refresh your mind as to the relative positions of these parts.

Begin the dissection of the following muscles on the left side of your cat. In this manual, when reference is made to the **"right" or the**

"left" side, it always means **to the right or the left side of the cat** and does not mean to your right or left as you look at the illustration. Think of your own body as being in the same position as that of the cat and you can determine accurately which is its right or left side. Remember that all drawings were made from large, well-developed cats. If you have a small specimen, you will have more difficulties. Dissect each muscle so it can be demonstrated to the instructor or to anyone else.

1. The **cleidocervicalis** arises from the clavicle and can be identified as follows. The **cephalic vein** appears as a dark streak passing over the shoulder and turns deeply into the base of the neck. This vein will not show if the blood has been drained out. Begin at the place where this vein turns into the neck and separate or loosen the caudal edge of the cleidocervicalis muscle dorsal to and covering the vein above the shoulder. Insert a probe under the caudal edge of this muscle and toward the lower jaw. Find the cranial edge of the muscle and loosen its sides from the head to the clavicle and from the underlying muscles. The lower cranial edge of the muscle covers part of the large external jugular vein, previously seen. Many muscle

fibers pass over the clavicle and continue down the forelimb, where they form the cleidobrachialis muscle. The cleidocervicalis inserts on the nuchal crest of the skull and the first few cervical vertebrae. The dorsal half of the caudal edge of this muscle close to the vertebrae is continuous with or fused with the cervical trapezius muscle immediately caudal to it. Separate these two muscles with scissors to the middorsal line. The lower portions of these muscles are separated by the **omotransversarius**, most of which lies medial to the cleidocervicalis. Bisect (cut in half) the cleidocervicalis muscle at right angles to the direction of its fibers and reflect the two ends. This muscle draws the clavicle forward and upward or the head and neck ventrally.

2. The **cleidobrachialis** is a continuation of the cleidocervicalis below the clavicle and down the cranial part of the brachium, or upper arm. At its lower end it merges with the superficial pectoral muscle of the chest. Its caudal edge is close to the cephalic vein. Loosen it along this edge, pull it up, and discover its cranial limit as it merges with the superficial pectoral muscle along a white line. Separate along this line, bisect, and reflect. This muscle compares to one of three that represent the deltoideus of man. The other two are the **spinodeltoideus** and **acromiodeltoideus**, which are discussed later in this group.

3. The **trapezius**, pars **cervicalis**, covers the upper part of the scapula, and most of its cranial edge merges with the cleidocervicalis. It arises from fascia along the middorsal line near the tips of the neural spines of the thoracic vertebrae, and it inserts on the spine of the scapula. Near its cranioventral edge the cervical trapezius is jointed by the **omotransversarius**, which attaches to the acromion of the scapula. The muscle fibers of the cervical trapezius have receded from the neural spines, leaving the muscular sheath that serves as a tendon. Bisect through the muscular portion slightly below the upper edge of the scapula. This muscle helps to hold the upper edges of the scapula against the body.

4. The **trapezius**, pars **thoracalis**, is caudal to and continuous with the cervical trapezius and originates from the tips of the neural spines of the caudal thoracic vertebrae. It inserts on the caudal portion of the spine of the scapula, and the fibers extend to the upper lateral edge of the scapula. (In man these three muscles are merged into one large trapezius.) The thoracic trapezius forms the lower portion of the large trapezius, which pulls the scapula toward the spinal column and dorsalward. Loosen the lower caudal edge of the thoracic trapezius of the cat, insert a probe under its entire width, and bisect at right angles to the muscle fibers. Be sure to reflect this muscle completely because it covers part of the following muscle.

5. The **latissimus dorsi** is large, with its upper cervical edge covered by the thoracic trapezius. It arises in fascia and aponeurosis from the opposite side along the middorsal line of the caudal thoracic region and most of the lumbar region. It covers most of the lateral surface of the body and extends forward and downward below the scapula where a portion of the cutaneus trunci may be seen. The latissimus dorsi inserts on the shaft of the humerus, but before doing so it merges with the following muscles: (a) the **teres major** from the ventral edge of the scapula, (b) the **tensor fasciae antebrachii** from the inner surface of the upper arm, and (c) the deep pectoral muscles. Do not try to identify these muscles at this time. Loosen the lower and upper edges of the latissimus dorsi along its entire extent and insert a probe under the middle, bisect, and reflect. The spinal nerves appear as white threads coming to this muscle to its medial surface. The number of these nerves indicates the number of myomeres that have merged to form the muscle. At the upper caudal limits this muscle becomes continuous with a thin sheath that serves as a tendon. Apparently the muscle fibers have receded from the neural spines. (This muscle is attached to the humerus and gives the cat much power in pulling backward when running or climbing.) The amount of the cutaneus trunci muscle that remains on the lower part of the latissimus dorsi depends on the size of the muscle and the amount left on the skin when the skin was removed.

6. The **deltoideus**, pars **scapularis** (**spinodeltoideus**) is covered by thin fascia near the cephalic end and below the spine of the scapula. Cut off this fascia but do not injure the cephalic or circumflex veins. The spinodeltoideus is about one-half inch wide and is immediately ventral to the spine of the scapula. Its lower edge is almost lateral to the lower edge of the scapula. The origin of the spinodeltoideus is in fascia.

The lower end continues as a wide tendinous sheath that extends under the cephalic vein and the next muscle to be considered and inserts on the humerus. It helps to bend and outwardly rotate the humerus. Insert a probe under it, bisect, and reflect.

7. The **deltoideus,** pars **acromialis** (**acromiodeltoideus**) is about the same size as the spinodeltoideus. Sometimes it appears double as it originates from the glenoid border of the acromion process of the scapula and at times on the adjacent suprahamate process. The greater part inserts on the tendon of the spinodeltoideus, but the outer fibers continue to the humerus. In both cat and man this muscle flexes the humerus and rotates it outwardly. Bisect and reflect, separating it from underlying muscles.

The **cephalic vein** gives off a branch below and caudal to the lower end of the spinodeltoideus. This is the **caudal circumflex,** which usually passes deeply below the spinodeltoideus to join the axillary vein (see Fig. 4-1). The cephalic vein usually continues up over the shoulder, turns inward between the cleidocervicalis and omotransversarius muscles, and joins

the subscapular vein, which empties into the axillary vein. A superficial branch does not turn deeply but continues across the outer surface of the cleidocervicalis and joins the external jugular vein directly. Color the cephalic vein blue.

Cats vary greatly in the completeness, or thoroughness, in which their blood vessels are injected. Some blood vessels are well injected in some cats, whereas these same vessels are not injected in other specimens. If they are not injected, you probably cannot identify them.

DEEP MUSCLES OF NECK AND SHOULDER (Fig. 2-2)

In Fig. 2-2 the superficial muscles have been removed.

1. The **splenius** is large and flat, covering most of the side of the neck close to the vertebrae and medial to the rhomboideus capitis. It is composed of many short muscles that are united into one mass, and it aids in raising the head. This muscle cannot be loosened or bisected with satisfaction.

2. The **rhomboideus capitis** lies under the

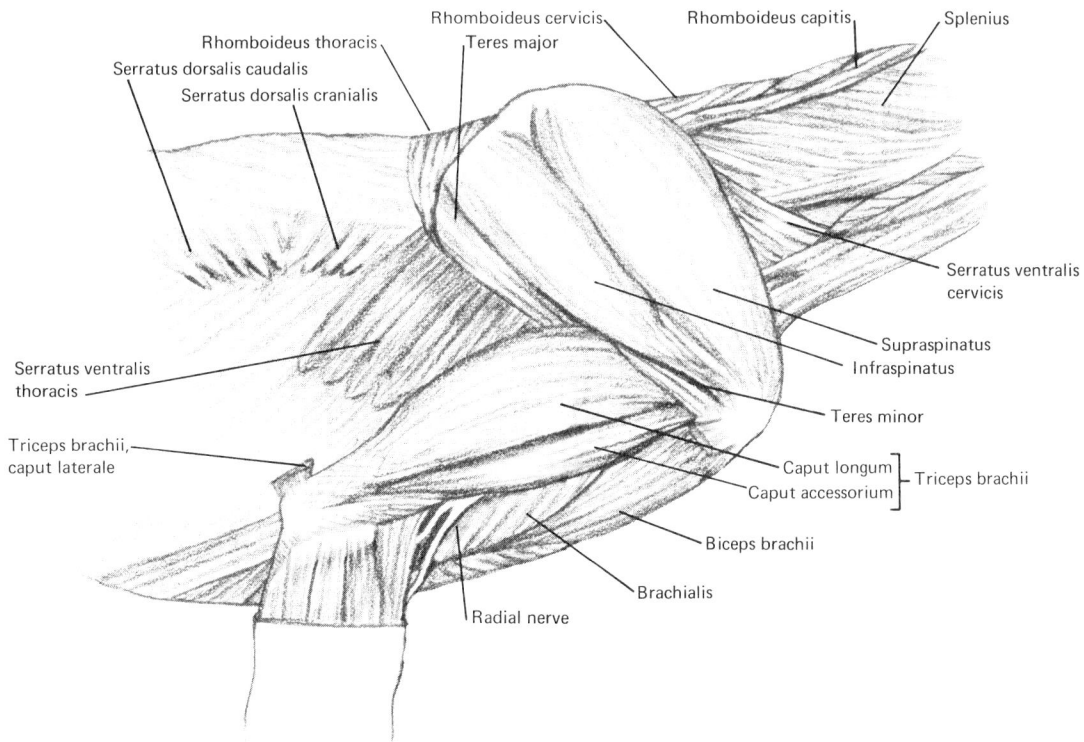

Fig. 2-2. Deep muscles of neck, shoulder, arm, and thorax.

cleidocervicalis on the upper lateral part of the neck. It is attached to the nuchal crest and inserts on the upper edge of the scapula. It is less than a centimeter wide and lies close upon other neck muscles. In man this muscle is a part of the rhomboideus minor. Its action is to rotate and draw the scapula forward. Bisect and reflect.

3. The **rhomboideus cervicis (minor)**, medial to the cervical trapezius, is larger than the rhomboideus thoracis (major) in the cat and lies cranial to it. Its origin is on the spinous process of the cervical and thoracic vertebrae and extends to the dorsal border of the scapula. The insertion is caudal to and in close contact with the insertion of the occipitoscapularis muscle. It draws the scapula dorsalward and forward. Pull the upper edge of the scapula away from the thoracic vertebrae about two inches and bisect carefully, leaving half attached to the scapula.

4. The **rhomboideus thoracis (major)** lies along the caudal edge of the rhomboideus minor, and their edges are often partly fused. It arises from the thoracic neural spines and inserts at the dorsocaudal angle of the scapula, where it becomes more distinct. Bisect carefully, leaving half attached to the scapula. The rhomboideus major and minor seem to be incorrectly named on the cat, since the major is the smaller; however, they were first named on man, where the major is the larger.

5. The **omotransversarius** (Fig. 2-1) arises from the transverse processes of the cervical vertebrae under the cranial end of the cleidocervicalis and extends backward, where it joins the cranial edge of the cervical trapezius, and the two muscles insert on the lower portion of the spine of the scapula. The omotransversarius draws the scapula forward. Bisect. See the mounted skeleton for the relationship of bones. This muscle is not found in man.

6. The **cleidomastoideus** lies deep to the lower edge of the cleidocervicalis on the ventrolateral region of the neck. It arises on the clavicle and inserts on the mastoid process. The clavicle is very small in the cat and can be located most quickly by feel. The cleidomastoideus is very closely associated with the sternomastoideus along its medial edge. The external jugular vein extends across the lateral surface of the sternomastoideus but not across the cleidomastoideus. In man these two muscles are usually merged with one another and are called sternocleidomastoideus, but occasionally they are similar to the cat. **Do not bisect** the cleidomastoideus.

7. The **serratus ventralis thoracis (serratus anterior)** arises on several ribs as separate myotomes and extends forward and dorsalward, passing between the scapula and the chest wall to insert on the inner upper edge of the scapula. In the upper portion, medial to the scapula, the myotomes are merged into a compact muscle, with little indication of its myotome units. This muscle was first named in man, where the body is in an upright position and the muscle is really in an anterior position. The myotome origin on the ribs is serrated, which means notched. In the cat this muscle pulls the scapula caudally and downward. **Do not bisect this muscle.**

8. The **serratus ventralis cervicis (levator scapula)** lies immediately in front of the serratus ventralis thoracis and is continuous with it, so that the two muscles may appear as one. Pull the upper edge of the scapula away from the thoracic wall and see that these two muscles are continuous with one another. The transverse colli artery, a branch of the costocervical artery, comes through the body wall and near the line of fusion. The serratus ventralis cervicis may be seen medial to the lower part of the omotransversarius, previously identified. The serratus ventralis cervicis pulls the scapula forward and toward the sternum. **Do not bisect this muscle,** since it and the serratus ventralis thoracis are left to hold the leg to the body.

9. The **supraspinatus** covers the craniolateral surface of scapula, which is the supraspinatus fossa. It arises along the whole surface of the supraspinatus fossa. Its fibers converge to a point at the shoulder, pass over the shoulder joint, and insert on the **greater tubercle of the humerus**, cranial to the acromiodeltoideus. **Do not dissect or bisect.**

10. The **infraspinatus** occupies the infraspinatus fossa on the outer side of the scapula caudal to the spine. It arises from the fossa and spine of the scapula, the acromion, and the suprahamate processes and converges into a tendon that is attached to the outside of the greater tubercle of the humerus. **Do not dissect or bisect.**

11. The **teres major** lies along the lower caudal edge of the scapula and passes forward and downward, medial to the muscles of the upper arm. It merges with the latissimus dorsi. It can be loosened from the lower half of the

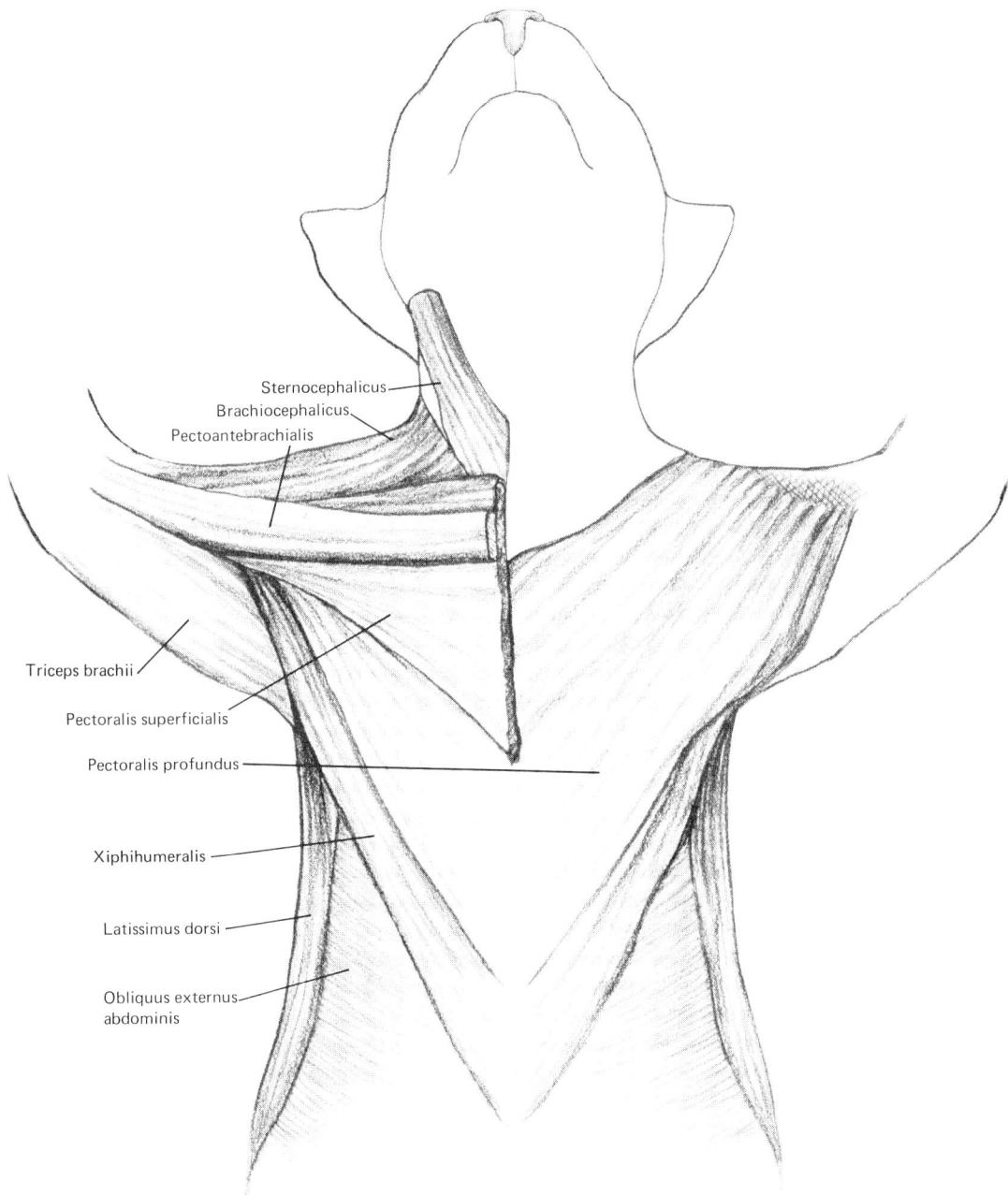

Sternocephalicus
Brachiocephalicus
Pectoantebrachialis

Triceps brachii

Pectoralis superficialis

Pectoralis profundus

Xiphihumeralis

Latissimus dorsi

Obliquus externus abdominis

Fig. 2-3. Pectoral muscles, ventral view.

edge of the scapula close to the point at which it merges with the latissimus dorsi and other muscles. Bisect and reflect the lower half only.

12. The **teres minor.** Reflect the lower half of the spinodeltoideus and upper half of the acromiodeltoideus; then the teres minor may be separated from what appears to be a part of the caudal edge of the infraspinatus. It arises from the caudal edge of the scapula, medial to the upper portion of the acromiodeltoideus, and

inserts on the greater tubercle of the humerus below the infraspinatus. **Do not bisect.** Remember that these muscles should not be mutilated but should be left so they may be identified again later.

VENTRAL VIEW OF PECTORAL MUSCLES (Fig. 2-3)

Lay the cat on its back in a dissecting tray. Spread the legs apart and tie them securely.

37

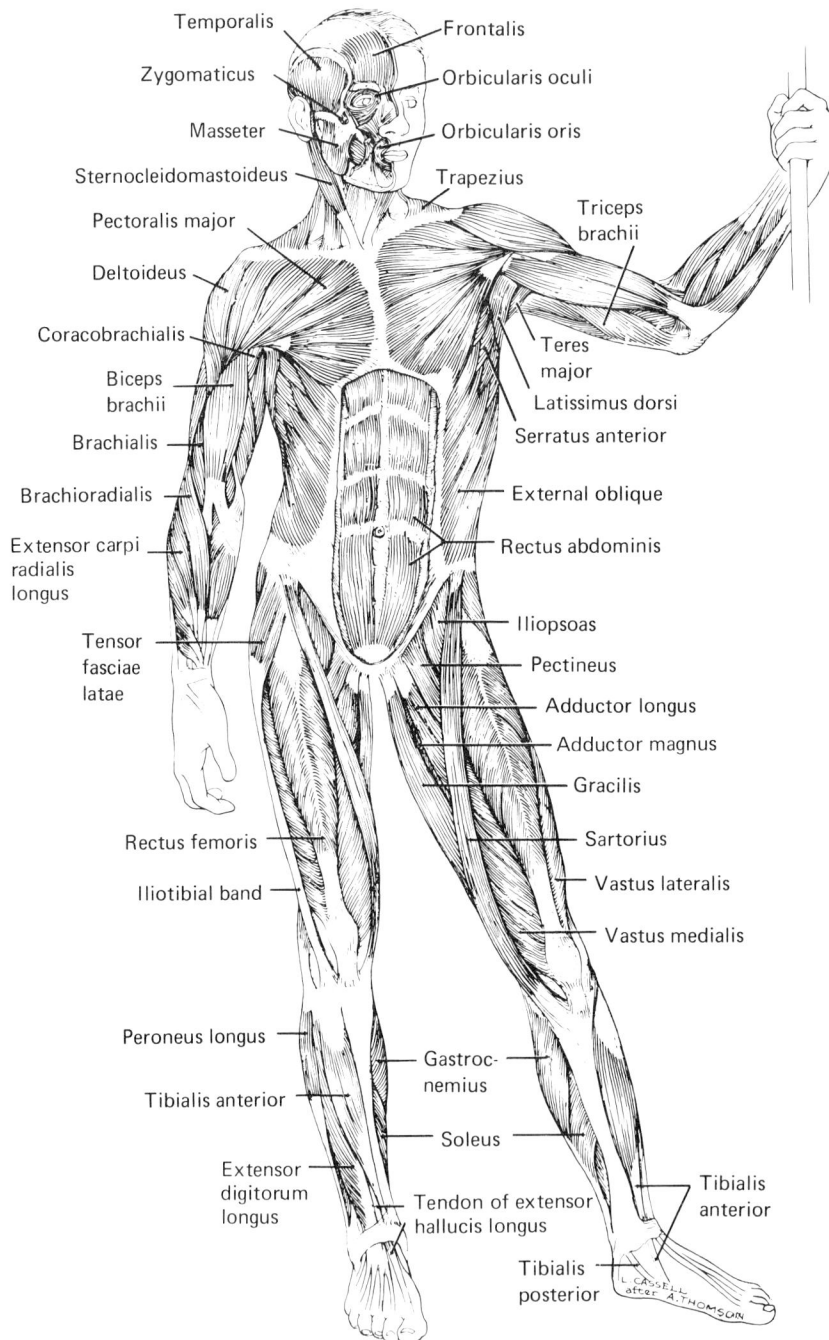

Fig. 2-4. Muscles of the human body, anterior view. (From Millard, N. D., King, B. G., and Showers, M. J.: Human anatomy and physiology, Philadelphia, 1956, W. B. Saunders Co.)

Now you are ready to dissect and study the **pectoral muscles.** They arise on the bones of the **sternum** and extend laterally to the **humerus** of each front leg. Examine the mounted skeleton of the cat to refresh your memory of the exact locations of the **manubrium, sternebrae,** and **xiphoid portions** of the sternum.

As you dissect the pectoral muscles of the left chest, you will find that they do not separate or dissect from one another as easily as most of the preceding muscles. There are two layers of pectoral muscles in the cat. Each is somewhat divided.

1. The **pectoantebrachialis** is really a half-inch-wide strip of the pectorales superficiales. It arises from the caudal part of the manubrium, passes laterally tight against the surface of the pectoralis superficialis and merges with the cleidobrachialis at about the middle of the humerus. It continues as a flat tendon and inserts on the ulna. Action pulls the arm median-ward. Often, thin white lines indicate the edges. Loosen these edges and insert a probe under this thin, ribbonlike muscle, bisect, and reflect.

2. The **pectoralis superficialis (pectoralis major)** is two or three inches wide and lies immediately dorsal to the pectoantebrachialis. It is folded on itself and may appear as two or more parts. Its cranial edge is short, extending from the manubrium, ventral to the first rib and to the head of the humerus. It often merges with the sternomastoideus and cleidobrachialis. The origin is on the sternum from the manubrium to the fifth or seventh rib, and it inserts along most of the lateral surface of the humerus. The caudal edge may be determined by the difference in the direction of its fibers from those of the pectoralis profundus, which lies deep to it. Loosen the pectoralis superficialis along its cranial and caudal edges and insert a probe under its entire width—this is difficult because the muscle may be in several parts and folded upon itself. Bisect and reflect.

3. The **pectoralis profundus (pectoralis minor)** lies partly dorsal and caudal to the pectoralis superficialis, and its fibers extend obliquely. Its origin is on the sternum from the fourth to twelfth ribs; it inserts near the proximal end of the humerus, usually by thin fascia. Sometimes it divides into the cranial and caudal sections. The fibers of this muscle are loosely associated with one another, and along its caudal edge it is fused with the following muscle so completely that the line of separation is indefinite. Bisect the pectoralis profundus in a straight line at right angles to its fibers. Reflect its proximal end to the sternum and expose the cranial thoracic vein. On the lateral surface is the long thoracic vein; you should also see the external thoracic artery and second thoracic nerve entering the pectoral muscles.

4. The **xiphihumeralis** is really a caudal portion of the pectoralis profundus. Its fibers are parallel to and fuse with the caudal edge of the pectoralis profundus. It is identified from the fact that its fibers arise on the xiphoid process of the sternum. It inserts on the humerus, but its fibers merge so completely with the **pectoralis profundus** and **latissimus dorsi muscles** that its identity is lost.

The pectoral muscles of the cat are represented in man by the pectoralis major and pectoralis minor. When these are cut on the cat, the axillary artery and vein and also the nerves of the brachial plexus may be seen passing from the body to the limb. Dissect away the fat and other connective tissue so they may be clearly seen. The arteries are red and the veins blue. Compare the cat muscles with those of man (Fig. 2-4).

A comparison of the nerve supply of the pectoral muscles in cat and man seems to indicate that the pectoantebrachialis and xiphihumeralis of the cat have merged with the other two muscles in man. The general belief among anatomists when comparing two closely related animals having similar structures is that the one having the least number of parts is the more specialized.

Name_____

Date_____

REVIEW QUESTIONS ON MUSCLES OF NECK AND THORAX

1. How does one determine the origin and the insertion of a muscle?

2. Name five muscles that insert on the scapula.

3. Name five muscles that originate on the scapula.

4. Name three muscles that merge with the latissimus dorsi.

5. Which muscle attached to the scapula shows its myotomes, or muscle plates, most clearly?

6. Name three muscles that arise at the middorsal line near the thoracic neural spines.

7. Name three skeletal muscles that show progressive fusion of myotomes.

8. Where do the mammary glands lie in reference to the cutaneous trunci muscle? (See the dissected specimen of a lactating cat.)

9. State the origin and the insertion of the serratus ventralis cervicis and of the omotransversarius muscles.

10. Name two locations where smooth, or nonstriated, muscles are found.

11. State the origin and insertion of each of the two principal cutaneous muscles of the cat.

12. What is the position of the cephalic vein in reference to the shoulder muscles?

13. Name the muscles that attach the forelimb and the pectoral girdle to the body.

14. What is peculiar about the naming of the rhomboideus muscles in the cat? (Examine your specimen and read the directions.)

15. What is the advantage to the cat in having the pectoral girdle attached by muscles to the body and spinal column?

16. What is the advantage to the cat in having the pelvic girdle attached directly to the bones of the sacrum?

17. Define a "raphe" and give two examples. (See definitions of terms.)

18. Define a "symphysis" and give two examples. (See definitions of terms.)

19. Name the four principal kinds of skeletal muscles in the cat and man and state the characteristic of each.

20. What is the general conclusion usually formed after considering comparable structures in closely related or in different species of vertebrate animals?

21. What is the principal difference in the trapezius muscles of the cat and man?

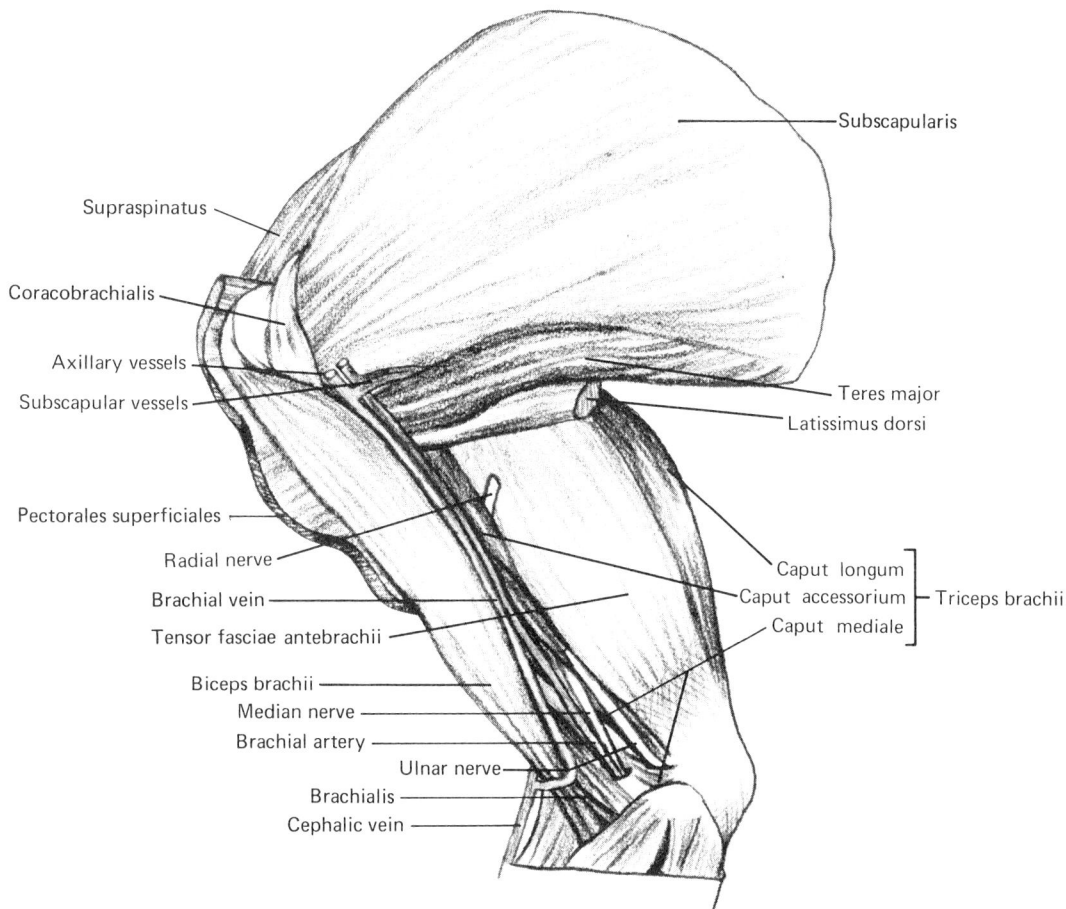

Fig. 2-5. Muscles of shoulder and arm, medial view.

CAUDAL MUSCLES OF BRACHIUM, OR UPPER ARM (Fig. 2-5)

1. The **triceps brachii** consists of several muscles on the caudal and lateral surfaces of the humerus. They are separated into three portions: caput laterale, caput longum, and caput mediale.

a. The **caput laterale** or **lateral head** (Fig. 2-7), is on the lateral surface of the upper arm immediately caudal to the cleidobrachialis and brachialis but is much larger and extends the entire length of the humerus. Its caudal edge can be easily separated from the muscle caudal to it. Loosen the edges of the caput laterale along the lateral surface of the upper arm from its origin on the upper surface of the humerus to the insertion on the ulna. Bisect and reflect. Observe the circumflex humeral artery along the medial surface of the upper half of the reflected part of the caput laterale.

b. The **caput longum** is about twice as large as the previous muscle and lies along the caudal surface of the upper arm. It originates medial to the spinodeltoideus and teres major from the caudal border of the scapula, and it inserts on the **olecranon process.** Reflect the upper portion of the spinodeltoideus to find the upper cranial edge. Before bisecting, loosen the tensor fasciae antebrachii muscle, which is on the medial surface of the upper arm. The tensor fasciae antebrachii is quite thin and flat on the caudal half of the medial surface, and its upper end merges with the latissimus dorsi. Bisect the muscle so that the medial surface of the caput longum may be seen. Now bisect the caput longum.

c. The **caput mediale,** or **medial head,** consists of three parts:

(1) the **first part** (often listed as **caput accessorium,** or **accessory head,** of triceps brachii, such that the combined muscle consists of four heads in the cat) arises from the caudal upper third of the humerus, cranial to the **caput longum,** passes downward, and joins the second

43

part of the caput mediale about two-thirds the distance to the olecranon process. The **radial nerve** passes between the first and second parts of the medial head of the triceps brachii and extends out on the lateral surface of the second part of the medial head. It soon divides into the long **superficial radial nerve** to the forearm and a branch to the muscles of the elbow. **Do not bisect this part of the muscle.**

(2) The **second part** of the medial head of the triceps muscle, which arises from about the middle third of the caudal surface of the **humerus,** lies parallel to the first part, and together they pass to the upper surface of the olecranon process.

(3) The **third part** of the medial head of the triceps brachii muscle is about one inch in length and lies on the inner or medial surface of the lower end of the humerus. It cannot be seen from the lateral surface. This part should not be confused with the anconeus muscle, which is on the lateral surface of the humerus. Therefore, **turn your cat so that its head will be to your right and its back toward you. Then turn the left front limb up over the cat's back** to get the medial view of the limb. If you have a left humerus, examine its supracondylar foramen as you read the following description and place the humerus of the cat in its proper position. In order to find the **supracondylar foramen** reflect the lower portion of the tensor fasciae antebrachii, previously bisected, and dissect the connective tissue from the supracondylar foramen and the inner surface of the olecranon process. Identify the **brachial artery,** which accompanies the **median nerve** as they pass through the supracondylar foramen. Insert a small probe into this foramen. Now we should be able to identify this small muscle. This **third part of the caput mediale muscle** arises from the part of the humerus that forms the arch of the supracondylar foramen, passes caudally, and inserts on the inner surface of the **olecranon process.** Each of these parts of the triceps helps to extend the elbow.

When the three heads of the triceps brachii muscle contract, they straighten out the front limb quickly, enabling the cat to force the thorax and head up. At about the same time the muscles of the hind limbs give the propelling power for the long jump.

2. We come now to the second main division of the muscles of the brachium, or upper arm.

The **tensor fasciae antebrachii** should have been bisected when the parts of the triceps were located. It is an exceptionally thin muscle on the medial side of the upper arm. It arises on the fascia of the lateral surface of the ventral border of the latissimus dorsi and inserts by a broad tendon onto the olecranon process. Sometimes it merges with the fascia of the pectoralis superficialis. It is not found in man. It extends the elbow.

3. The **anconeus** is easily confused with the third part of the medial head of the triceps brachii muscle, which is on the inner, or medial, side of the arm. The anconeus, however, is on the outer, or lateral, side of the lower part of the humerus (Fig. 2-6) and inserts on the lower edge of the semilunar notch of the ulna and thus supports the elbow joint. Turn your specimen so that you get the lateral view of the left front limb and identify the anconeus distal to the second part of the medial head of the triceps. It is an extensor of the elbow.

MEDIAL MUSCLES OF SHOULDER AND CRANIAL MUSCLES OF UPPER ARM (Fig. 2-5)

Again turn your cat around so that its head is to your right and its back toward you. Now turn its left limb up over the back toward you so you can more easily examine the medial, or inner, surface of the scapula and brachium of the left forelimb.

1. The **subscapularis** fills the entire subscapular fossa on the medial side of the scapula. Its origin is the entire subscapular fossa, except in the area of attachment of the serratus ventralis. Its insertion is by a flat tendon into the dorsal border of the lesser tubercle of the humerus. The teres major muscle, previously identified, lies ventral to the lower edge of the scapula. Ventral to the teres major is a part of the cranial end of the latissimus dorsi. These two muscles come forward medial to the long head of the triceps and insert on the humerus. They are the same in man. The subscapularis acts as an adductor of the humerus.

2. The **coracobrachialis,** about one-half inch long, is on the inner side of the shoulder joint and covers the insertion of the subscapularis muscle. It arises by a rounded tendon from the apex of the coracoid process. Fleshy fibers insert on the medial surface of the humerus near its proximal end and above the insertion of the

teres major. It is the same in man. It acts as an adductor of the humerus. **Do not bisect.**

3. The **biceps brachii** is large and spindle shaped, lying on the cranial surface of the humerus, medial to the insertion of the pectoralis superficialis. Do not confuse it with the brachialis. The biceps brachii arises by a strong tendon from the supraglenoid tubercle proximal to the glenoid cavity of the scapula. The tendon passes through the intertubercular groove of the humerus. It is the same in man, except that the muscle arises by two heads. It acts as the flexor of the forearm (elbow) and a supinator, since it turns the arm so the palm is upward. Bisect.

4. The **brachialis** arises on the lateral surface of the upper part of the humerus and inserts on the medial surface of the ulna. Reflect the lower half of the acromiodeltoideus and dissect off the fascia below its insertion. The brachialis lies lateral to the insertion of the pectoralis superficialis close against the lateral surface of the humerus. It crosses the cranial surface of the elbow and inserts on the medial surface of the ulna below the semilunar notch. It is a flexor of the elbow. It is the same in man. **Do not bisect.**

Some of the principal arteries, veins, and nerves may be identified at this time (Fig. 2-5).

1. **Arteries** (see also Fig. 5-1). Observe the **axillary artery** as it comes through the thoracic wall to supply the shoulder and forelimb. Do not cut the serratus ventralis muscles, since they should be left to hold the limb to the body. The axillary artery gives off the external thoracic branch to the pectoral muscles immediately lateral to the body wall. The **subscapular artery** gives off a branch passing between the subscapular and the teres major muscles. The main artery, now known as the **brachial**, passes medial to the humerus, where it gives off the **deep brachial artery** to the triceps brachii muscle and further on a muscular branch, the **superficial collateral artery**, to the biceps brachii muscle and the radial side of the forearm. As the brachial artery approaches the elbow joint, it gives off the **superficial brachial artery** and the **collateral ulnar artery**, which passes medial to the olecranon process. The main artery is then called the **median** and passes through the supracondylar foramen to descend in the limb.

2. **Veins** (see also Fig. 4-1). The **axillary vein** enters the thoracic wall below or caudal to the **external jugular** and continues as the **subclavian** under the clavicle. The main tributary to the axillary vein is the brachial.

The **brachial vein** lies on the medial side of the upper arm, or brachium, and receives a small vein, the **profundus brachii,** from the biceps muscle, usually coming between the teres major and the latissimus dorsi. In the lower portion of the upper arm the brachial vein is formed by the union of the **radial** and **ulnar veins.** The **axillary vein** receives the **external thoracic** and also the **subscapular,** usually immediately lateral to the thoracic wall, but the entrance position of the latter varies considerably in different cats. Trace the subscapular vein to where it comes from between the teres major and the latissimus dorsi.

The large **external jugular vein** comes down the ventrolateral surface of the neck and enters the thorax anterior to the axillary vein. The external jugular lies almost immediately under the skin and receives the **transverse scapular vein** a few centimeters before it enters the thorax. The transverse scapular vein usually receives the **cephalic** from the lateral surface of the shoulder and upper arm (Fig. 2-1), which was identified when beginning the dissection of the muscles. The **caudal circumflex vein** branches from the cephalic in the region of the acromiodeltoideus and passes in deeply to connect with the **subscapular vein.** The **caudal circumflex vein** can be seen caudal and on a level with the head of the humerus between the insertions of the teres major and latissimus dorsi muscles, where it becomes continuous with the subscapular or joins the brachial, which is a tributary of the axillary vein (Fig. 2-5).

In man all blood from the cephalic vein passes by way of the **posterior circumflex** to the subscapular, and none enters the transverse scapular.

Unless the veins are well injected, you will probably be unable to identify them. Occasionally a marked variation from the normal in position or structure is found; this is an anomaly. The size of the vein when injected depends largely on the amount of material and pressure used during injection. These vessels will be considered again when the venous system as a whole is studied.

3. **Nerves** (see also Fig. 8-4). The **median** (Figs. 2-5 and 2-8) is the easiest nerve to identify, since it has three long roots that unite almost at the same level as the head of the humerus;

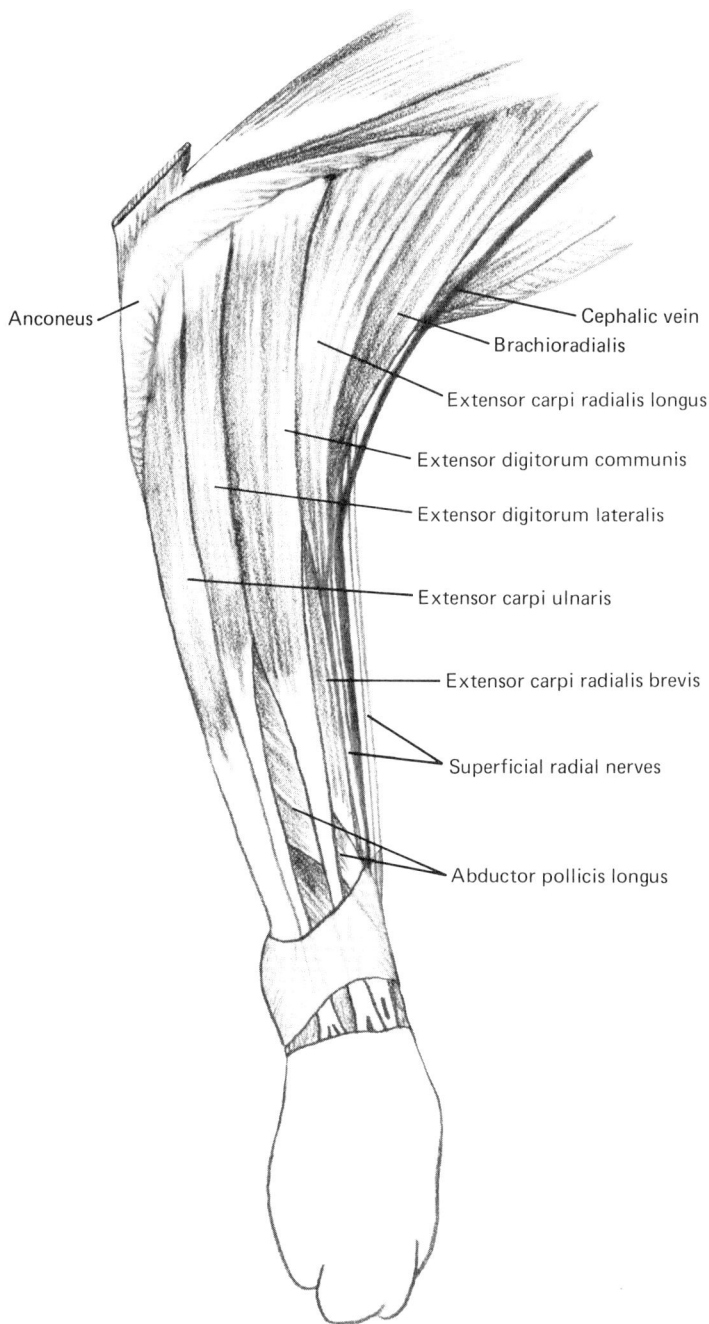

Anconeus

Cephalic vein

Brachioradialis

Extensor carpi radialis longus

Extensor digitorum communis

Extensor digitorum lateralis

Extensor carpi ulnaris

Extensor carpi radialis brevis

Superficial radial nerves

Abductor pollicis longus

Fig. 2-6. Superficial muscles of forearm, craniolateral view.

trace it until it passes through the supracondylar foramen of the humerus with the median artery. The **ulnar** (Figs. 2-5 and 2-8) is the most caudal of the three largest nerves and becomes superficial and is easily exposed against the medial condyle of the humerus; it then continues over the olecranon. The **radial** (Figs. 2-5 and 2-6) extends diagonally across the humerus and

divides into the **superficial radial** and the **muscular branches.** Color the blood vessels and add arrows to show the direction of flow.

SUPERFICIAL LATERAL MUSCLES OF FOREARM (Fig. 2-6)

Examine the lateral surface of the left forearm. The lower part of the humerus, at the place where the **ulnar nerve** is "close" to the inner surface of the elbow, is the "crazy bone." Dissect off several layers of connective tissue, or fascia, from the lateral surface of the forearm, being careful not to injure the underlying muscles. These muscles are so much alike that it is difficult to remember them accurately unless they are studied in some definite order. Therefore, we will take them in order, beginning with the one most cranial and proceeding to the one most caudal, as seen from the lateral surface. Identify each muscle at or near its origin on the humerus near the edge of the **anconeus** and trace it distally.

1. The **brachioradialis** muscle is on the lateral surface of the humerus below the lower end of the acromiodeltoideus, medial to the **superficial radial nerve** but lateral to the brachialis. It gradually decreases in size as it extends down the craniolateral surface of the radius and parallels the superficial radial nerve. It inserts on the styloid process of the radius. The **brachiocephalic vein** passes up the cranial portion of the limb and the brachioradialis muscle. Color the veins blue and the nerves green in the drawing. Bissect but do not damage the vein.

2. The **extensor carpi radialis longus** has a broad origin and is larger and distal to the **brachioradialis,** but it has about the same general proportions. It extends down the lateral surface of the radius. Its distal third is a small tendon inserted on the dorsal surface of the second metacarpal. It lies against and is often fused with the brevis muscle, discussed next. Its function is to help extend the paw. Bisect distally to the former muscle.

3. The **extensor carpi radialis brevis** is usually partly covered by the previous muscle, particularly at the upper end. It is often fused with the previous muscle, longus, but it ends in a slender, short tendon on the third metacarpal. The **deep branch** of the radial nerve passes under the head of this muscle. This muscle helps to extend the paw. **Do not bisect.**

The following three muscles originate on the humerus near the **anconeus,** previously considered in the group of lateral brachial muscles.

4. The **extensor digitorum communis** is exposed laterally for the entire length down the lateral surface of the arm; the middle third merges gradually into a slender muscular tendon. The tendon divides at the wrist and attaches to the second, third, fourth, and fifth digits, which

it extends. Dissect out and bisect at a different level than that for former muscles.

5. The **extensor digitorum lateralis** (corresponds to the extensor digiti quinti proprius of man) has about the same appearance as the former muscle, communis, but it is usually slightly broader, lies lateral to the ulna, and divides to each of the third, fourth, and fifth digits. It extends these digits. Bisect distal to that of number four.

6. The **extensor carpi ulnaris** originates from the lateral epicondyle of the humerus and extends down the arm lateral to the region separating the radius and ulna. It is rather uniform in size and often has a shiny tendon. It inserts on the tubercle on the ulnar side of the base of the fifth metacarpal. Bisect distal to that of number five.

7. The **extensor digiti primi et secundi** (Fig. 2-7) is small and arises immediately caudal to the origin of the extensor carpi ulnaris on the craniolateral surface of the ulna by short, fleshy fibers below and lateral to the semilunar notch. Fibers extend distally and toward the radius, tight against the ulna. The lower half has a white, shiny tendon. Much of this muscle lies medial to the extensor carpi ulnaris and supplies the first and second digits.

These muscles have long tendons attached to the bones of the toes. Tendons differ from ligaments in that they extend from muscles to bones, while ligaments extend from bone to bone.

If the muscles are becoming dry, place damp paper towels or cloth under the skin before wrapping and putting away at the end of each laboratory period.

DEEP LATERAL MUSCLES OF FOREARM (Fig. 2-7)

Now reflect the proximal ends of the preceding muscles (two to five) and find the interosseous nerve, which extends laterally to the origin of the following muscles.

1. The **supinator** is flat, slightly spiral, and extends down the lateral and cranial surfaces on the proximal end of the radius. It arises by ligaments from the lateral surface of the humerus and the upper outer surface of the radius. It passes distally and medially, converges, covers the upper surface of the radius, and inserts on the proximal third of the ventral surface of the radius. **Do not bisect.**

2. The **abductor pollicis longus** (abductor

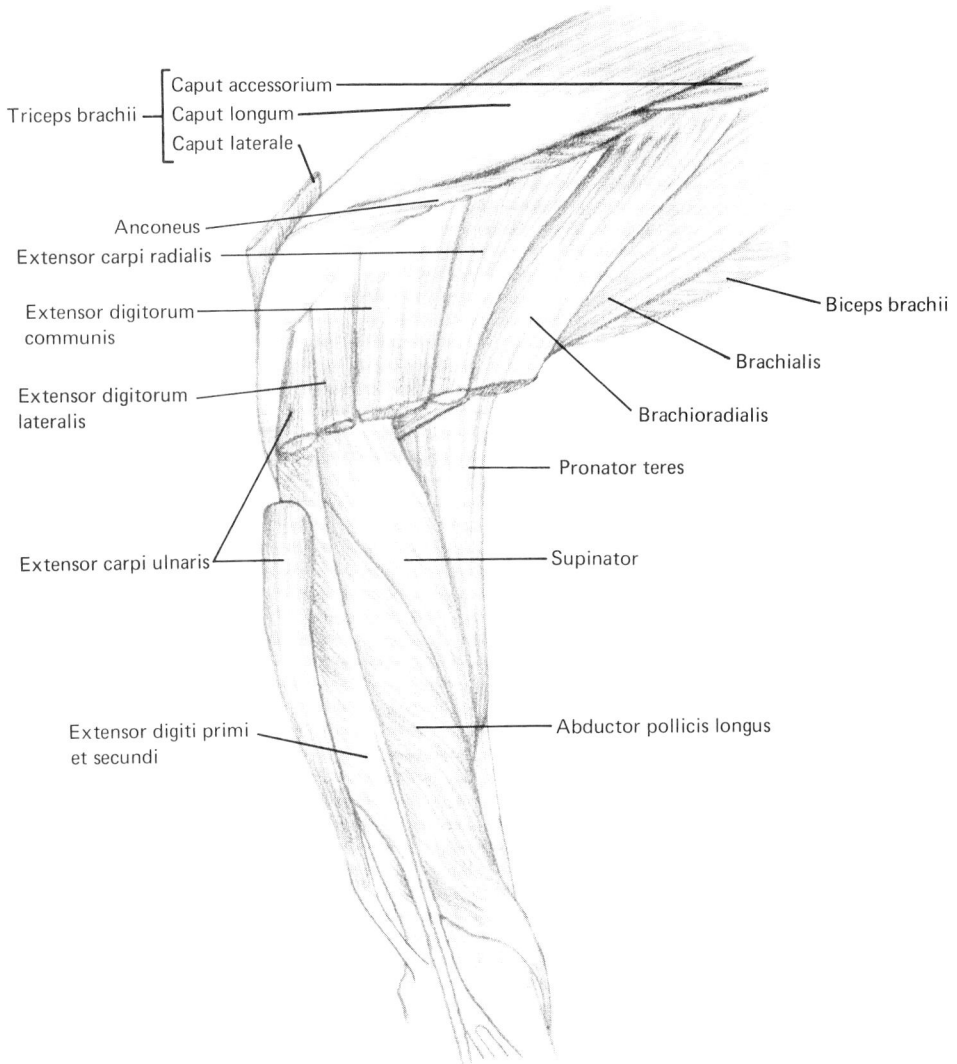

Fig. 2-7. Deep muscles of forearm, craniolateral view.

digiti primi longus) extends from the distal cranial surface of the ulna and from the caudal surface of the lateral third of the radius. Fibers converge to form a flat tendon that passes obliquely over tendons of the **extensor carpi radialis longus** and **brevis** on the craniomedial surface of the forearm above the carpus. **Do not bisect.**

SUPERFICIAL MEDIAL MUSCLES OF FOREARM (Fig. 2-8)

Turn the cat so that its head is to your right and its back is toward you. Turn the left forelimb up over the back; cut off the heavy fascia from the medial, or inner, side of the left fore-

limb. The **brachioradialis** and the **extensor carpi radialis brevis** are shown on the inner side of the bend, with their origins on the lateral surface of the humerus. These muscles were identified in the previous exercise but should be recognized from this view so that they may not be confused with others.

1. The **pronator teres** arises from the medial epicondyle of the humerus near the lower end of the bony arch forming the supracondylar foramen. It passes distally, parallel with and caudal to the median artery, which is a continuation of the brachial artery. It decreases definitely in size, and its lower half may be difficult to trace to its insertion on the radius. **Do not bisect.**

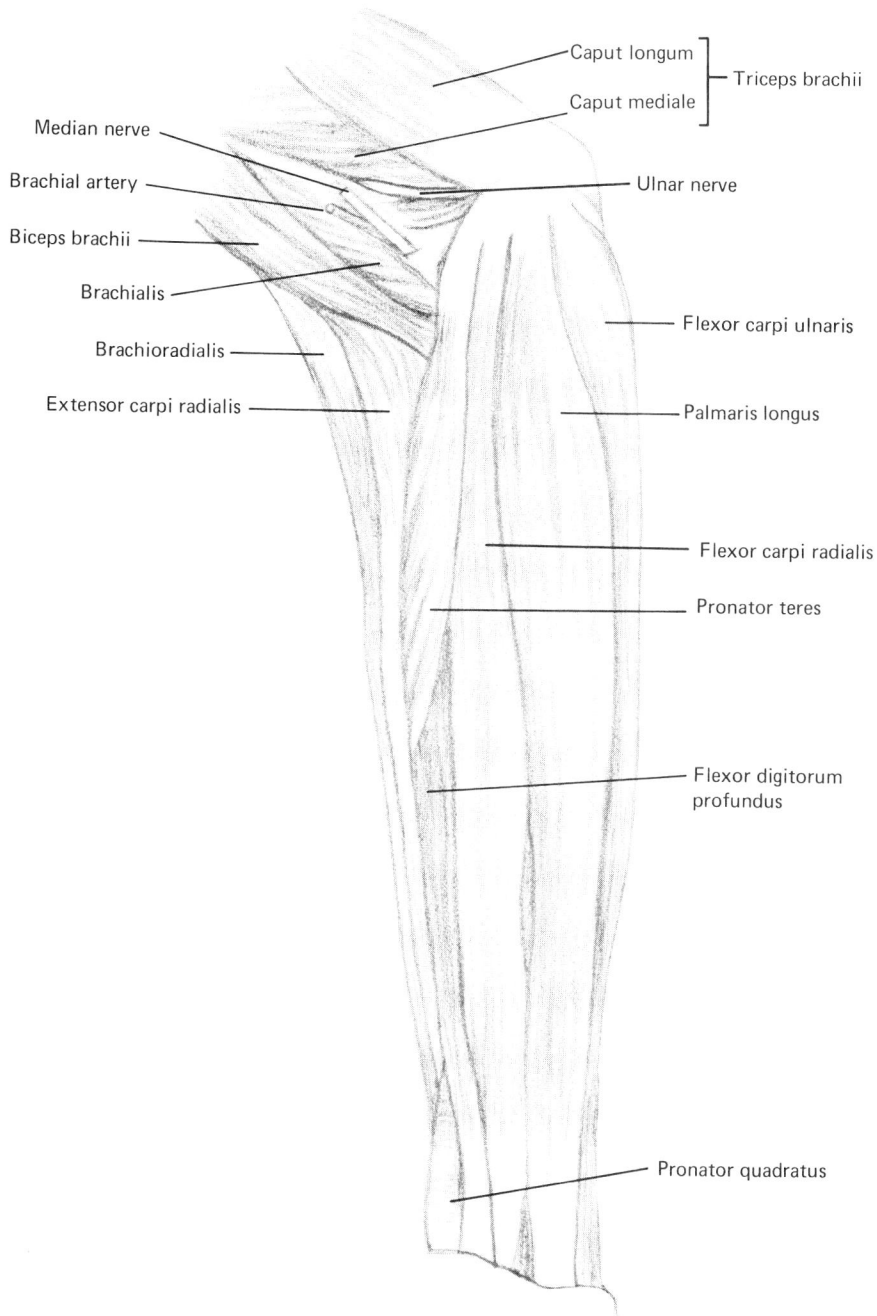

Caput longum ⎤
Caput mediale ⎦ Triceps brachii

Median nerve

Brachial artery

Biceps brachii

Brachialis

Brachioradialis

Extensor carpi radialis

Ulnar nerve

Flexor carpi ulnaris

Palmaris longus

Flexor carpi radialis

Pronator teres

Flexor digitorum profundus

Pronator quadratus

Fig. 2-8. Superficial muscles of forearm, caudomedial view.

2. The **extensor carpi radialis brevis** is partially seen between the supinator and the abductor pollicis longus. It extends the carpus.

3. The **flexor carpi radialis** originates on the medial epicondyle of the humerus, distal to the origin of the previous muscle; it extends distally to a more cranial position at the lower end of the radius, gradually tapering into a small tendon lateral to the median artery. The outer cranial surface is against the pronator teres, and the caudal surface is against the flexor digitorum profundus and sometimes the palmaris longus. The median artery often lies on the lower half of this muscle. Bisect and reflect.

4. The **palmaris longus** is a large, flat, broad muscle in the center of the medial surface. It arises from the epicondyle of the humerus and is considered to be a superficial portion of the

49

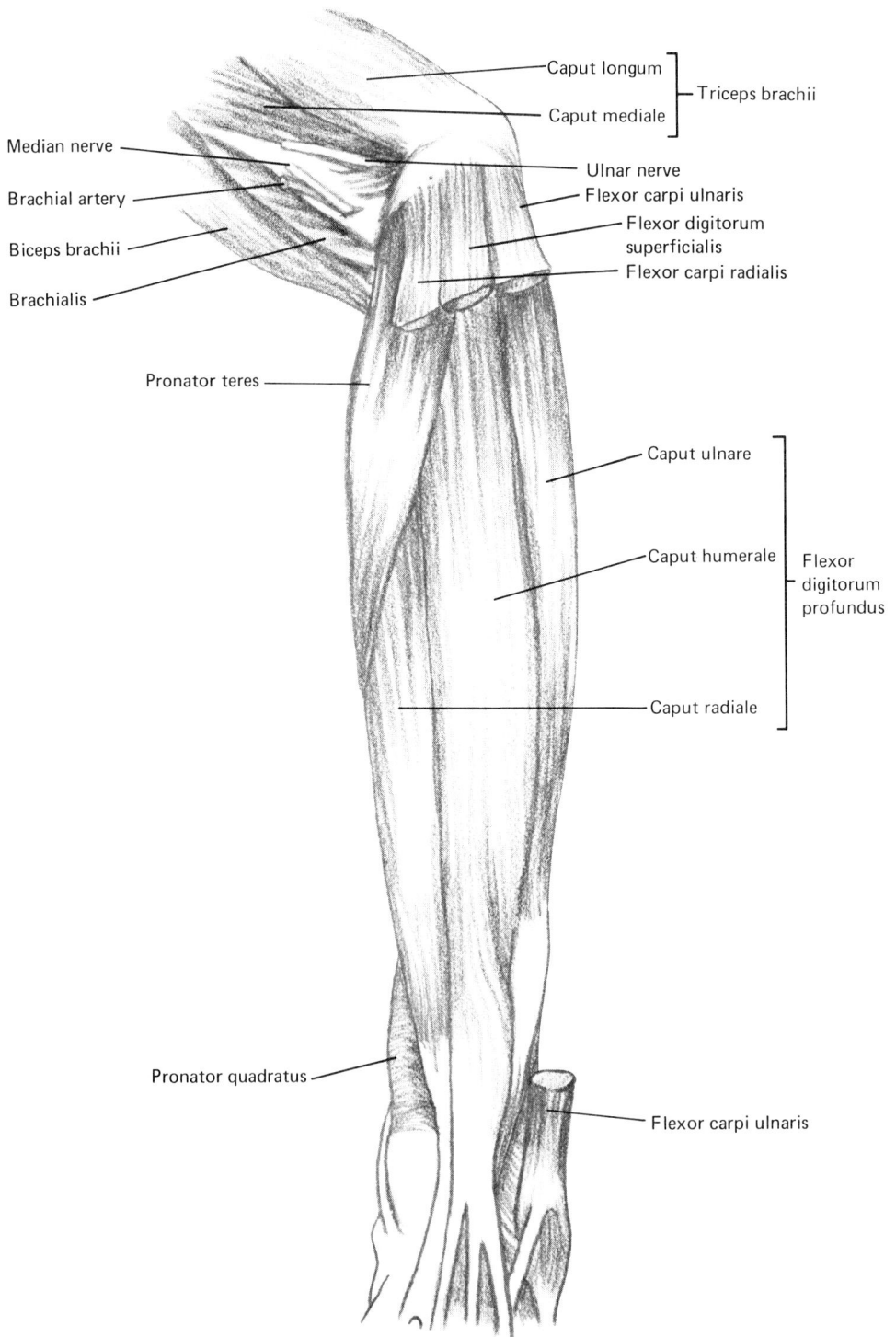

Fig. 2-9. Deep muscles of forearm, caudomedial view.

flexor digitorum superficialis. It ends in a flat tendon at the carpus, passes through the **transverse ligament,** divides into several tendons, and spreads out into a pad in the palm of the hand, which divides, giving off tendons to the digits. Bisect and reflect.

5. The **flexor digitorum superficialis** has an **ulnar** and a **radial part.** Cut and reflect the **transverse ligament** at the carpus and dissect out the lower end of the palmaris longus muscle, which is really a part of this muscle. The ulnar part of the **flexor digitorum superficialis** originates from the lower lateral surface of the palmaris longus. The radial part arises from the area of merging of the tendons from the first and second parts of the flexor digitorum profundus but soon divides and supplies the second and third digits. **Do not bisect.** This muscle is much larger in man.

6. The **flexor carpi ulnaris** appears to have two heads. One arises on the medial epicondyle of the humerus, the other on the medial part of the olecranon process of the ulna. The two heads may be separated by the **ulnar nerve** and pass down as separate muscles almost to the carpus, where they unite into one tendon to be inserted onto the bones of the carpus. They act to bend or flex the carpus. Bisect and reflect, cutting through the lower third.

DEEP MEDIAL MUSCLES OF FOREARM (FLEXOR DIGITORUM PROFUNDUS GROUP) (Fig. 2-9)

Keep the specimen in the same position as when dissecting the former group of muscles.

The superficial muscles are reflected to expose the **flexor digitorum profundus group of muscles.** This group is called **one muscle** but appears as **five parts,** or **heads.** (According to nomenclature listing, there are only three heads: **ulnar, humeral,** and **radial;** the second, third, and fourth heads are really divisions of the humeral head.) These five muscles are more or less fused with one another, but this is different in each cat. Hence, the following description may not be entirely accurate for a given animal. These muscles are considered from the ulna forward toward the radius. They function in flexing the digits.

1. The **first head** (ulnar) is under the caudal head of the **flexor carpi ulnaris** on the medio-

caudal surface of the ulna. It extends distally and is closely attached to the medial surface of the ulna. **Do not loosen or bisect.**

2. The **second head** arises from the distal end of the medial epicondyle of the humerus cranial to the **ulnar nerve,** which separates the first and second heads of the flexor digitorum profundus down to the carpus. Loosen the nerve and see where it passes over the elbow. The **flexor carpi ulnaris** covers most of the first two heads. The second, third, and fourth heads, divisions of the humeral head, are more or less fused with one another. The way they fuse with one another varies in different cats. Do not force their separation. Bisect the second head and find the fourth head, which is usually directly under it.

3. The **third head** arises from the medial epicondyle of the humerus, under the heads of the **flexor digitorum superficialis** and **flexor carpi radialis.** It lies caudal to and, throughout most of its length, in contact with the second head. Its proximal portion merges with the fourth head, but distal parts are separate before reaching the carpus. **Do not bisect.**

4. The **fourth head** arises from the medial epicondyle of the humerus under the second head, with which it is closely associated or fused halfway to the carpus. The distal portion usually decreases in size abruptly into a small definite tendon. In some cases the fourth head is fused with the first, but their distal ends can usually be separated. **Do not force the separation of these heads.**

5. The **fifth head** (radial) arises from the upper third of the radius, and its fibers are in contact with the **pronator teres** under the third head. The fifth head joins the third at the carpus, where it is probably easiest identified. If you have a small cat, the fifth head may be difficult to identify. It lies close against the medial surface of the ulna and forms a flat, muscular tendon between the lower ends of the ulna and radius. The radial artery and vein and branches of the median nerve are conspicuous along the medial side of the radius.

In man the flexor digitorum profundus group is a single large muscle covering the ulnar side of the forearm. It arises mostly from the upper ventral and medial surfaces of the ulna and ends in four tendons supplying the second to the fifth digits.

The **pronator quadratus** is not a part of the flexor digitorum profundus group. It is seen under the fifth head, however. Its fibers extend from the lower end of the ulna distally and forward to join the radius distal to those of the fifth head. Its fibers are deeper and extend at an oblique angle to those of the radial head.

REVIEW QUESTIONS ON MUSCLES OF BRACHIUM AND ANTEBRACHIUM

1. What is the advantage to the cat in having the triceps muscles larger and stronger than the biceps and brachialis?

2. Name two large muscles attached on the olecranon process of the ulna.

3. Name the muscles inserted on the humerus c' se to the biceps and the brachial muscles.

4. Name two small muscles that originate on the distal end of the humerus and insert on the ulna.

5. Name the two largest muscles inserted on the humerus.

6. What structures pass through the supracondylar foramen?

7. How many muscles or parts constitute the flexor digitorum profundus group of muscles?

8. What is the difference between a tendon and a ligament? (See definition of terms.)

9. What advantage is there in having long, narrow tendons extending over the carpal and meta-carpal bones to the various digits?

wo small muscles with origins on the scapula and insertions on the humerus.

11. Name three large muscles between the scapula and the ribs.

12. Name a small muscle of the shoulder that inserts on the humerus.

13. Name two small muscles of the elbow that originate on the humerus.

14. What constitutes the so-called "crazy bone"?

15. What is the olecranon process, and what is its function in relation to movements of the fore-arm? (Examine your specimen.)

16. Where are the principal blood vessels located in reference to the upper arm and humerus? (See your specimen.)

17. What is the general function of the flexor digitorum profundus group of muscles?

18. Where are the muscles that control the principal movements of the toes located? (Examine your specimen.)

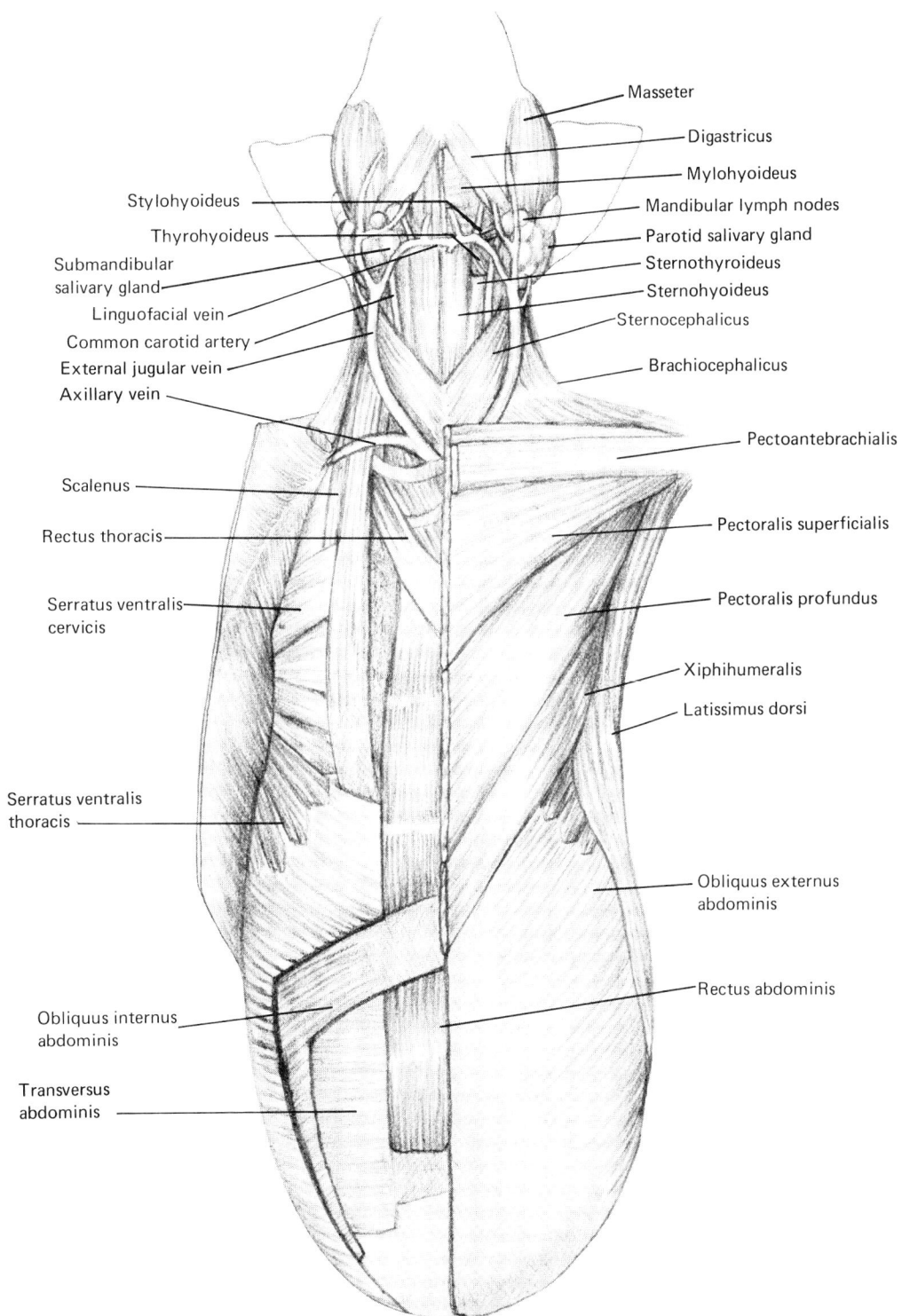

Fig. 2-10. Ventral muscles of neck, lower jaw, and thorax.

Masseter

Digastricus

Mylohyoideus

Mandibular lymph nodes

Stylohyoideus

Thyrohyoideus

Parotid salivary gland

Submandibular salivary gland

Sternothyroideus

Sternohyoideus

Linguofacial vein

Sternocephalicus

Common carotid artery

External jugular vein

Brachiocephalicus

Axillary vein

Pectoantebrachialis

Scalenus

Pectoralis superficialis

Rectus thoracis

Pectoralis profundus

Serratus ventralis cervicis

Xiphihumeralis

Latissimus dorsi

Serratus ventralis thoracis

Obliquus externus abdominis

Rectus abdominis

Obliquus internus abdominis

Transversus abdominis

VENTRAL MUSCLES OF NECK, LOWER JAW, AND THORAX (Fig. 2-10)

Pass a probe under the pectoral muscles; bisect, if not previously done, and reflect to expose the smaller muscles close to the ribs.

1. The **rectus thoracis (transverse costarum)** and its tendons extend over the first five ribs. It extends from the sternum at about the base of the fifth rib diagonally toward the clavicle but inserts on the first and second ribs. It assists in bracing the chest wall. Loosen its edges and bisect. This muscle is absent in man.

2. The **scalenus** is long and slender; it extends lateral to the rectus thoracis but also ventral to most of the thorax from the ninth rib forward dorsal to the **external jugular vein** to attach to the transverse processes of the cervical vertebrae. Loosen the edges of this vein and also those of the sternomastoideus, finding the myotomes going to some of the ribs. **Do not bisect.** Observe the tributaries of the **external jugular vein** as follows (see also Fig. 4-1): (a) The **hyoid arch (transverse jugular) vein** extends across near the **hyoid cartilage** or bone to join the external jugular on the opposite side. (b) The **linguofacial (rostral facial) vein** arises over the lateral edge of the mouth. (c) The **maxillary (caudal facial) vein** enters the external jugular near the base of the ear, ventral to the **submandibular salivary gland.** The **rostral** and **caudal auricular** tributaries usually unite with one another below the **parotid salivary gland,** which is below the ear. Two usually large **lymph nodes** lie close to the linguofacial vein, and one lies caudal to the submandibular salivary gland. All lymph nodes eventually drain by capillary tubules into the lymphatic system. The **sublingual salivary gland** is small and deeply located immediately rostral to and in touch with the submandibular salivary gland. The **internal maxillary artery** passes close to it.

3. The **sternohyoideus** arises on the manubrium under the sternomastoideus, passes forward close to the median line over the cricoid and thyroid cartilages, and inserts on the hyoid cartilage. Separate the two thin sternohyoid muscles along the median line, loosen the right one, bisect, reflect its cranial half, and find the sternothyroideus immediately under it, as seen from ventral view.

4. The **sternothyroideus** is shorter than the previous muscle, since it extends from the manubrium only to the thyroid cartilage. These last two named muscles help pull the larynx down when swallowing and are practically the same in man.

5. The **thyrohyoideus** extends from the lateral surface of the thyroid to the hyoid cartilage and is almost in line with the previous muscle. It is only about one-half inch long. Carefully dissect away the fascia, pass a probe under it, but **do not bisect.** Under the sternothyroideus is the trachea with its transverse cartilaginous rings. Rub these with your finger to help identify them. Lateral to the trachea and larynx is the **common carotid artery** which is usually injected with red latex. Lateral to the carotid artery is the vagosympathetic nerve trunk. Separate the larger vagus nerve from the sympathetic nerve on the right side of the cat. These nerves are shown in Fig. 8-3. The **thyroid gland** lies between the carotid artery and the upper end of the trachea and cricoid cartilage. It is cigar shaped, is reddish in color, similar to muscular tissue, and is held in place by loose connective tissue. Loosen the thyroid on the right side, but do not remove it. It is one of the endocrine glands, and its hormone is drained away by blood. In man an abnormality of this gland is known as goiter.

6. Three small muscles, whose fibers contribute to the formation of the tongue, can now be identified as follows: (a) The **genioglossus** arises near the symphysis, extends parallel to the geniohyoideus for a short distance, then passes dorsal to some small muscles, and inserts on the base of the hyoid cartilage. (b) The **hypoglossus** extends obliquely immediately caudal to the origin of the genioglossus and the center of the mandible and inserts on the base of the hyoid cartilage. (c) The **styloglossus** lies dorsal and medial to the digastricus, whose fibers are almost parallel with the lower jaw. These three muscles are extrinsic to the tongue, but they also pass into it and assist in its movements and in the elevation of the larynx. They are much the same in man.

7. The **masseter** is thick at the caudal outer surface of the lower jaw. It arises from the zygomatic arch and inserts on the mandible at the ventral border of the masseteric fossa. It is the same in man. It elevates the lower jaw. **Do not bisect.**

8. The **temporalis** is fan shaped and covers the temporal region of the skull in front of the ear and close to the temporal bone. It arises from

the squamous portion of the temporal bone and inserts on the lateral and medial surfaces of the coronoid process of the mandible. It helps elevate the lower jaw. **Do not bisect.** The hyoid, thyroid, and cricoid cartilages, as well as the embryonic cartilaginous jaws, and also the malleus, incus, and stapes arise from embryonic branchial arches.

BRANCHIAL MUSCLES (Fig. 2-10)

This group of muscles consists of those that develop from the branchial arches and are supplied by visceral nerve fibers.

1. The **sternocephalicus** arises on the manubrium and courses craniolaterally deep to the external jugular vein. The right and left muscles diverge toward their insertions on the skull. Each muscle divides near its insertion into sternomastoideus and sterno-occipitalis portions.

2. The **brachiocephalicus** (cleidocervicalis, cleidobrachialis, and cleidomastoideus) are described with the superficial and deep muscles of the neck and shoulder, pp. 33, 34, and 36.

3. The **trapezius** are described with the superficial muscles of the neck and shoulder, p. 34.

4. The **omotransversarius** are described with the deep muscles of the neck and shoulder, p. 36.

5. The **digastricus** lies close against the medial surface of the lower jaw. It arises on the jugular and mastoid processes of the skull and inserts on the rostral half of the mandible and forward to the symphysis. The **linguofacial vein,** a tributary of the external jugular, crosses the digastricus at the angle of the jaw and there receives the submandibular labial vein. This vein passes between the **digastricus** and the jawbone. Pass a probe under the digastricus at the angle of the jaw and bisect and reflect the rostral end to expose the **mylohyoideus,** whose fibers pass transversely from one jaw to the other. Reflect the caudal end of the digastricus far enough to expose the **right external carotid artery** from which arises the lingual artery. The **hypoglossal nerve** passes forward almost parallel to this artery and dorsal to the caudal edge of the mylohyoideus. The **axillary vein** joins the **external jugular** near the thoracic wall and unites with the **subclavian vein.** At the base of the neck the **transverse scapular** joins the **external jugular.**

6. The **mylohyoideus** is thin and its fibers extend from the inner surface of each jaw to the median line where they meet the corresponding muscle from the opposite side along the raphe. Lift up its caudal edge on the right side, starting where the hypoglossal nerve passes dorsal to it. Also reflect the sheath of fascia dorsal to the mylohyoideus, which contains branches of the hypoglossal nerve.

7. The **stylohyoideus** is about two millimeters wide and two centimeters long. It extends laterally from the base of the hyoid cartilage, dorsal to the external jugular vein and digastric muscle, between it and the lower jawbone, to the tympanic bulla. Here it is attached to the long cornua, or greater wing, of the hyoid cartilage. The greater wing of the hyoid in the cat is homologous with the lesser wing in man. The **stylohyoideus** is easily damaged or overlooked. It helps to pull the larynx up when swallowing. It is similar in man except that it is attached on the styloid process of the temporal bone.

8. The muscles that close the jaw are included with the masseter temporalis and the pterygoids.

9. The muscles of the face, pharynx, and larynx are not discussed here.

VEINS, LYMPH NODES, AND SALIVARY GLANDS (Fig. 2-10)

1. The **external jugular vein** is formed by three main tributaries at the base of the skull (see Fig. 4-1).

a. The **hyoid arch** (**transverse jugular**) connects with the opposite side in the region of the hyoid cartilage.

b. The **maxillary** is formed by tributaries from the side of the head and the region at the base of the ear.

c. The **linguofacial** is from the upper and lower jaws. Dissect off the platysma muscle fibers and connective tissue to expose these veins.

2. The **lymph nodes** in this region vary in size and position. Usually there is one or two near the base of the linguofacial vein; there is also often one large lymph node lateral to the cartilages of the larynx. This one is the most variable.

3. **Salivary glands** include the parotid, submandibular, and sublingual.

a. The **parotid gland** lies ventral to the external acoustic meatus. It is large and irregular in shape. It is drained by the parotid duct, which passes forward laterally across the masseter muscle. It then turns medially and enters the mouth lateral to the upper third premolar tooth. Remove the connective tissue surrounding and covering the parotid gland.

b. The **submandibular gland** lies ventral to the parotid and usually in contact with it. It is smooth and more uniform in shape than the parotid. The maxillary vein crosses its ventral surface and the submandibular duct drains it. The submandibular duct is small and passes laterally to the digastricus, then dorsal to the mylohyoideus, and opens into the mouth by a large papilla near the frenulum, or median membrane, under the tongue.

c. The **sublingual gland** is small, elongated, of slightly different color, and deeply situated at the rostromedial edge of the submandibular gland. Do not confuse it with the large, more superficial lymph nodes. The sublingual has two ducts, which extend nearly parallel with an open into the mouth near the opening of the submandibular duct. These ducts are difficult to identify. Cut the linguofacial vein near its base and reflect it and the submandibular gland, to which the sublingual is attached. Now you should see the elongated sublingual gland; notice how it lies between the masseter and digastricus muscles.

Slit open the larynx so that you can definitely feel and see the three cartilages. Pass a probe into the slit of the larynx dorsally and forward through the glottis into the pharynx and mouth.

REVIEW QUESTIONS ON VENTRAL MUSCLES OF NECK AND LOWER JAW

1. Name the principal muscles that are attached to the lower jaw.

2. Which muscles are extrinsic to the larynx?

3. What is the principal difference between the cleidomastoideus and sternomastoideus in the cat and man?

4. What is the relationship of the position of the sternomastoideus to the external jugular vein?

5. Name a muscle that extends transversely between the lower jaws and also name a muscle in the region of the larynx.

6. Name the salivary glands and state their relative positions to one another.

7. State the exact location of the large lymph nodes that lie close to the salivary glands.

8. What two muscles lie parallel and ventral to most of the trachea?

9. Name seven branchial muscles.

10. Name three brachial muscles.

11. Name the principal muscles used in chewing.

12. What large vein is close to the surface in the neck?

13. Name two longitudinal muscles lying between the lower jaws.

14. What is the position of the thyroid gland in reference to the larynx and trachea?

15. Name the principal artery that carries the blood to the head. (Fig. 2-10)

16. Name three oblique muscles lying below the tongue and contributing to its movements.

17. Describe the shape and attachments of the scalenus muscle.

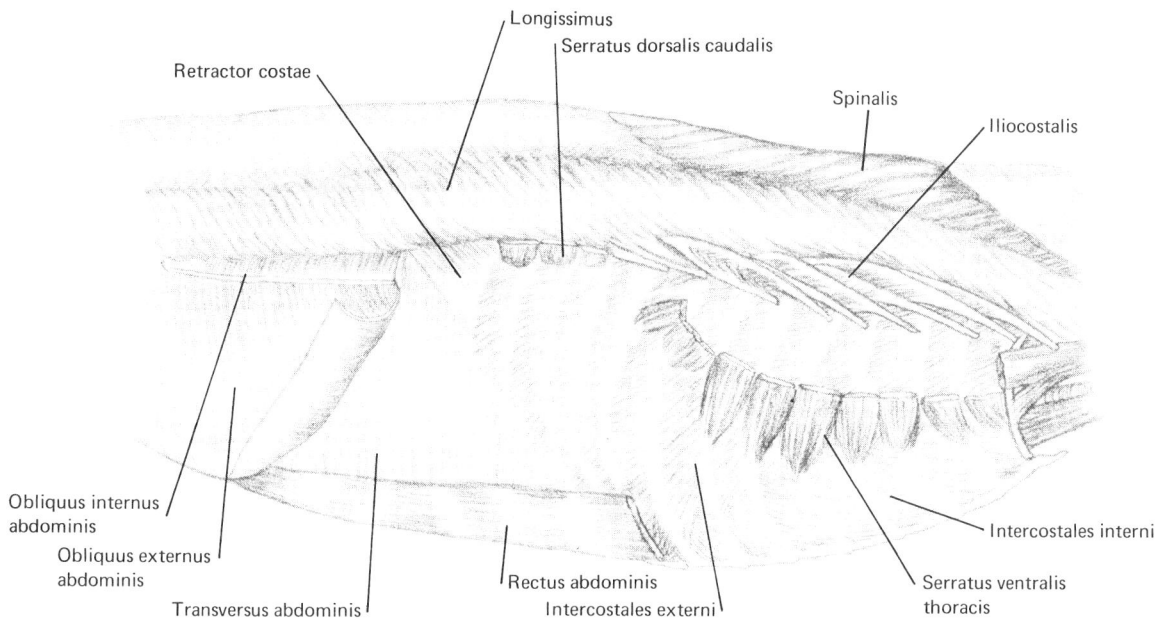

Fig. 2-11. Muscles of thoracic and abdominal wall, lateral view.

MUSCLES OF THORACIC AND ABDOMINAL WALL (Fig. 2-11)

1. The **serratus dorsalis cranialis** (see Fig. 2-2) may be identified as follows. Completely remove the upper portion of the latissimus dorsi muscle and reflect the trapezius and rhomboideus muscles as shown in Figs. 2-1 and 2-2. Find four or five rather indistinct myotomes above and caudal to the scapula, coming down from near the thoracic vertebrae and attaching to the ribs. Cut the common sheath covering them and with a dissecting needle determine the direction of the muscle fibers. This muscle was first named on man as posticus superior, where in the erect position it is truly above, or superior to, the following muscle, the serratus posticus inferior. These muscles help elevate the ribs.

2. The **serratus dorsalis caudalis** is much like the preceding muscle. It is thin, lying caudal to the serratus dorsalis cranialis and medial to the upper part of the latissimus dorsi. It arises by four or five fleshy heads, or myotomes, from the last four or five ribs. Passing dorsally the fibers unite to insert on the lumbar spinous processes and intervening interspinous ligaments. Bisect.

3. The **intercostales externi** may be identified as follows. Pull the upper edge of the scapula away from the spinal column and with a needle determine the direction of the fibers of the ex-

ternal intercostals between the ribs immediately above the attachment of the serratus ventralis to the body wall. The origin is on the caudal margin of the cranial ribs. The fibers run **caudally** and **ventrally** to insert on the cranial margins of the ribs. Numerous external intercostals are found in the outer portion of the intercostal spaces. They are the same in man. Their action is the protraction of the ribs.

4. The **intercostales interni** may be identified as follows. Cut off a small area of the thin external intercostal fibers between the adjacent ribs, and observe the internal intercostal fibers extending **downward** and **forward.** These two layers of intercostals are believed to have originated from a common layer that extended more nearly parallel with the ribs. The action of the internal intercostals is the retraction of the ribs. The intercostals can be found almost anyplace between the ribs.

5. The fibers of the **obliquus externus abdominis,** which was mentioned previously, run caudally and ventrally. The muscle is large and thin, covering the lateral wall of the abdomen and part of the thorax. It arises from the last nine or ten ribs by numerous tendons, which are interwoven. These fibers expand out in a fanlike fashion with some caudal fibers inserted onto the cranial border of the pubis. The cranial fibers run downward into a median raphe, while

median fibers join the **linea alba,** which is the median white line of the abdomen, by means of an aponeurosis. The muscle is the same in man. The contracting action compresses the abdominal viscera from near the middle of the sternum to near the lumbar vertebrae. Bisect the external oblique at right angles to the fibers at a distance of about six inches, if not previously done, and reflect.

6. The **obliquus internus abdominis** is deep to the external oblique and is tight against the transversus abdominis. It arises from the lumbar fascia common to it and the external oblique, and by an aponeurosis it is united to the linea alba. Its fibers extend ventrally and slightly cranially, almost at right angles to fibers of the external oblique. The caudal fibers pass externally to the **rectus abdominis.** The internal oblique muscle is the same in man. It acts as a compressor of the abdomen. Bisect the muscle fiber area and reflect.

7. The **transversus abdominis** is the most internal of the abdominal muscles. It is thin and arises from cartilages of the false ribs, from transverse processes of the lumbar vertebrae, and from the ventral border of the ilium. Its fibers are almost parallel with those of the external oblique. A portion of the internal oblique is removed in the drawing to show the transversus abdominis, which acts as a constrictor of the abdomen. Under the transversus abdominis is the shiny transparent lining of the abdominal cavity, the **parietal peritoneum.**

Observe that in each of the layers of oblique muscles the fibers extend in a different direction. This arrangement is similar to the direction of the grain of wood fibers in plywood, which gives the abdominal wall greater strength for the amount of material involved. It is the same in man.

8. The **rectus abdominis** is about one-half inch wide, lying on each side of the midventral line, or linea alba, of the abdomen. This muscle extends from the **rectus thoracis** (Fig. 2-10) to the pubic symphysis. Dissect out the left rectus abdominis and determine its relationship to other muscles and their fascia, particularly the external and internal oblique. Where are the limits of the rectus abdominis? This muscle supports the abdominal wall, especially during pregnancy. **Do not bisect.**

The abdominal wall consists of the last four named flat layers of muscles, with the parietal peritoneum, connective tissue, and the skin. Fat is usually deposited in the connective tissue layers so that the fat layers and the lean meat, or muscle layers, alternate. This is true in all fat mammals, and in the pig, when the body wall is sliced, it is recognized as bacon. The small pieces of cartilage often found in the edge of a slice of bacon are sections of costal cartilages from the lower ends of ribs.

In appendectomies on man, the surgeon usually cuts through each layer of abdominal muscle, parallel to its muscle fibers. This permits firmer stitches to be taken in closing the wound, a more rapid recovery, and a stronger abdominal wall.

9. The **erector spinae** is a group of epaxial muscles along almost the entire spinal column. They do not differ from one another as much as those in other regions; hence, they are more difficult to identify. These muscles are lateral to the neural spines and dorsal to the transverse processes, mostly in the lumbar and thoracic regions. The larger muscles in these two regions may be identified rather satisfactorily, but the smaller muscles and those in the transition region from lumbar to thoracic are much more complicated and difficult to identify.

The erector spinae group of muscles consists of many subdivisions and is quite complicated. Only the main divisions, which extend forward through the lumbar, thoracic, and cervical regions will be considered here.

a. In the **lumbar region** the **longissimus** may be identified as follows. Reflect the dorsal portions of the external and internal oblique muscles and their sheaths to the middorsal line of the lumbar region. This should uncover a heavy white sheath, or aponeurosis, the **thoracolumbar fascia.** Cut through this sheath close to the spines and peel it laterally to the point where it passes between the **inner** and **outer** muscles. The less heavy sheath of the **transversus abdominis** and sometimes the fascia of the **internal oblique** turn medially lateral to the epaxial muscles. Separate these muscles from the **hypaxial** muscles, which are below and lateral to the transverse processes of the lumbar vertebrae. The **inner longissimus** is separated from the next more lateral muscle, the **iliocostalis,** by the heavy aponeurosis mentioned previously. Next to the vertebral spines are the **multifidi** and **rotatores** (not shown).

Identify the various epaxial muscles, which

vary in their relative sizes in different cats. They support the small of the back and bend the body from side to side. The muscles of the **lumbar region of the cow** are the **choicest** and **most expensive cuts of meat** in the meat market. In man these muscles are important in maintaining the erect position and in giving strength to the back.

b. In the **thoracic** and **cervical regions** the **longissimus** and **iliocostalis** continue forward. You will find three main longitudinal muscle areas in the dorsal part of the thorax. We shall consider these, beginning with the most lateral.

(1) The **iliocostalis** has fibers that extend cranioventrally and unite by many small, white tendons to the upper parts of various ribs above the **serratus dorsalis cranialis** and **caudalis**.

(2) The **longissimus thoracis** is medial and dorsal to the iliocostales. This muscle is a con-tinuation of the same group seen in the lumbar region and may be traced in some specimens caudally, where the continuity is seen. However, it usually becomes quite complicated, and the dissection is seldom satisfactory. Trace the longissimus forward into the neck where, as the **longissimus cervicis**, it becomes closely associated with the **splenius**. The longissimus cervicis, divides into many small units that are attached to the cervical vertebrae.

(3) The **spinalis** is dorsal and medial to the longissimus in the thoracic region. It extends longitudinally lateral to and against the neural spines. Lateral to it on the dorsal surface of the cervical vertebrae lies the multifidus cervicis. **Do not bisect.** They are very much the same in man.

REVIEW QUESTIONS ON MUSCLES OF ABDOMINAL WALL

1. Name the muscles of the lateral abdominal wall.

2. In what direction do the muscle fibers of each of the abdominal wall muscles extend?

3. What is the advantage of the way in which abdominal muscle fibers lie in reference to each other?

4. Name four muscles of the lateral thoracic wall that show myotome, or myomere, arrangement.

5. How is adipose tissue, or fat, arranged in reference to the abdominal wall muscles?

6. Define and give an example of a retroperitoneal organ. (See definition of terms.)

7. In what direction are the abdominal muscles cut in an operation to remove the appendix? Why?

8. What is the name of the innermost layer of the abdominal wall?

9. In what direction do the fibers of the intercostal muscles run?

10. What is the linea alba and what is its significance? (See definitions of terms.)

11. Name three muscles that show myotomes quite clearly.

12. What muscles lie along the lumbar vertebrae?

13. What is meant by the term "aponeurosis"?

14. What muscles may be seen in a slice of bacon?

15. Sometimes round, hard objects that cannot be chewed are found in bacon. What are they?

16. How is the peritoneum arranged?

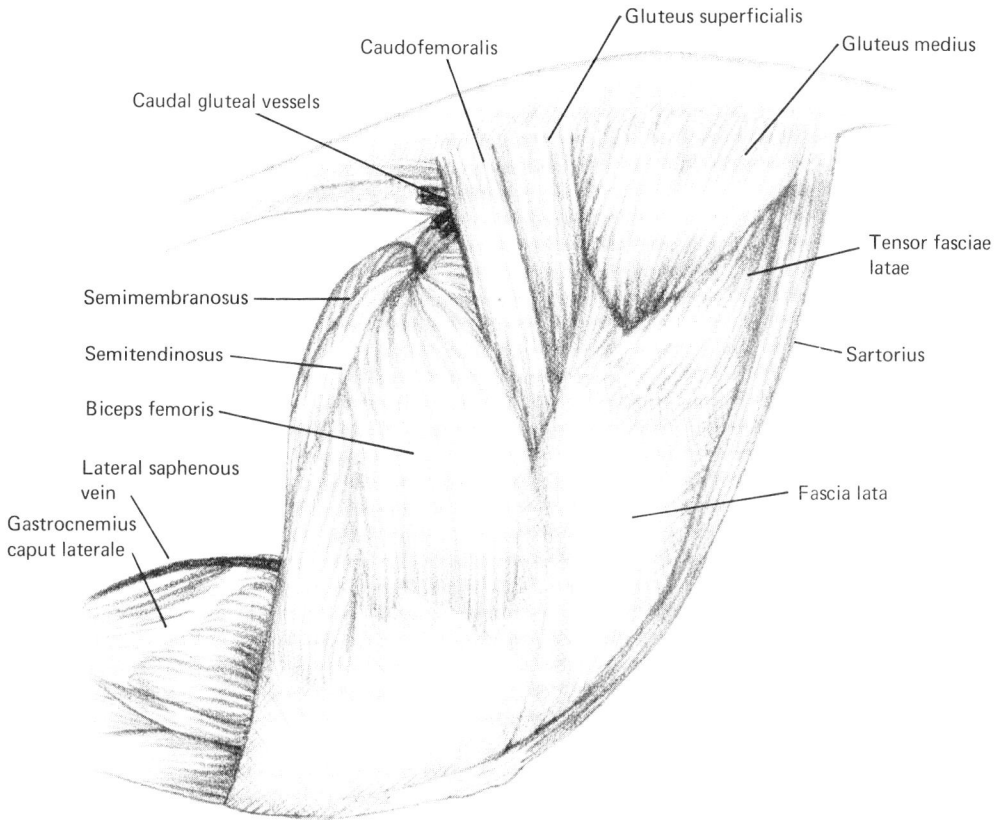

Fig. 2-12. Superfical muscles of hip and thigh, lateral view.

SUPERFICIAL MUSCLES OF HIP AND THIGH (Fig. 2-12)

Remove the fat and fascia from the cranial, medial, and lateral surfaces of the thigh, being careful not to injure any muscles. If the cat is a male, the sperm ducts, with their arteries and veins, will be found embedded in the fat near the pubic symphysis. The ductus deferens (sperm duct) is a dense white cord and appears superficially somewhat like a large nerve. Dissect out the sperm duct on one side from the testis through the inguinal canal, where it passes through the lower abdominal wall. The male reproductive organs will be considered more fully later.

1. The **sartorius** is about one and one-half inches wide and lies on the cranial portion of the medial side of the thigh. It often extends over the cranial surface and slightly to the lateral surface of the thigh. Compare this muscle as shown in Figs. 2-12 and 2-15. Loosen the lateral edges from the knee to the ilium and pass a probe under it from its origin on the ilium to its insertion by ligaments on the patella and internal tubercle of the tibia. It helps to pull the thigh toward the medial line and also rotates it. It is the same in man. After bisecting and reflecting, the circumflex iliac artery and vein may be seen supplying its undersurface.

2. Before dissecting the **biceps femoris**, observe the **lateral saphenous vein** along the caudal surface of the leg. This vein turns under the lower end of the biceps femoris toward the back of the knee through the fat to join the **femoral vein**, which may be seen later. A small **communicating branch** continues up the caudal surface of the thigh and connects with the **caudal gluteal vein** of the hip. If these veins are dilated with blood or are injected, you can see them easily.

The biceps femoris is large and powerful, covering much of the lateral surface of the thigh. It is two or two and one-half inches wide throughout most of its length. Find the cranial edge about one inch caudal to the sartorius, extending from the ischium to the tibia. Its caudal edge is approximately in line with that of the thigh and inserts on the upper half of the tibia. Find the caudal edge and pass a probe under it to its cranial edge. Loosen a small area, bisect the biceps femoris near its center, and reflect each of the ends completely. Much fat is usually deposited under its lower end. Pull out this fat with your fingers and expose the large **isciatic (sciatic) nerve** (see Fig. 2-14), which branches into the lateral **peroneal nerve** and the medial **tibial nerve.** The area that this fat occupies is the popliteal fossa, and through it the popliteal vein, previously mentioned, and artery pass. There is usually a lymph node embedded within the fat.

3. The **abductor cruris caudalis** (see Fig. 2-14) is often one-half centimeter wide and is sometimes difficult to locate. It is usually found on the medial side of the biceps femoris when reflected but extends almost parallel to the isciatic nerve. It arises from the transverse process of the second caudal vertebra. It extends along the inner caudal edge of the biceps femoris, with which it eventually fuses. It is absent in man.

4. The **caudofemoralis** is at the upper cranial edge of the biceps femoris and sometimes is attached to it. Its origin is on the transverse processes of the second and third caudal vertebrae. The upper half is muscular, whereas the lower half is simply the sheath of fascia, which is thin and small, parallel with and lateral to the femur. It is inserted into the middle of the lateral border of the patella. It is absent in man. It acts as an abductor of the thigh and a flexor of the shank. Bisect through its upper muscular portion on the level with the head of the femur. After bisecting you can see the base of the abductor cruris caudalis muscle better. Sometimes the caudofemoralis is reflected with the biceps femoris.

5. The **tensor fasciae latae** is a very peculiar muscle, thick and wide at its upper extremity but reduced to **heavy fascia (aponeurosis)** throughout its lower two thirds and merges with the aponeurosis of the caudofemoralis. Its origin is on the cranial end of the ilium and adjacent fascia, covering the craniolateral half of the thigh, inserting by long, broad fascia to the external surface of the femur and on the tibia. It is much reduced in man (Fig. 2-13). Locate the head of the femur and its entire extent by means of a dissecting needle. Cut the **fascia lata** where it joins the entire lateral surface of the femur. Pass a probe under this fascia and the muscular portion of this muscle and bisect through its thickest part. Separate the upper portion of the tensor fasciae latae from the femur and along the lower edge of the ilium.

Muscles are surrounded by fascia that sometimes thickens into a **heavy aponeurosis,** as is the case of the lower part of the tensor fasciae latae. The real muscle fibers have migrated or disappeared from this area, leaving only this heavy sheath, fascia lata. This has probably resulted in part from the fact that the area is struck by more objects when the animal runs.

6. The **semitendinosus** muscle is not half tendon, as the name implies, but is muscular, except near its insertion. It lies medial and caudal to the biceps femoris along the caudal portion of the thigh, is nearly uniform, and is about the size of a human finger. It arises from the tubercle of the ischium and inserts on the medial side of the tibia. It helps to flex the knee. Bisect and reflect.

7. The name of the **semimembranosus** is definitely misleading, since the muscle is large and muscular in the cat (see Figs. 2-12 to 2-16). It lies medial and cranial to the semitendinosus and the lower part of the **isciatic nerve.** This muscle must be observed from the medial surface of the thigh before it can be bisected properly. Identify the conspicuous lateral saphenous vein passing diagonally across the lower part of the thigh and the leg. Loosen the thin, broad gracilis muscle, bisect, and reflect it to expose the medial side of the semimembranosus. The semimembranosus muscle usually appears folded upon itself with two layers along its cranial edge and one along its caudal edge. Its cranial edge lies close to the adductor muscle. The insertion is along the distal caudal portion of the femur. It is a powerful muscle that helps to draw the leg backward.

The biceps femoris, semitendinosus, and the semimembranosus muscles are known as the "hamstring" muscles and are extensors of the hip and flexors of the knee.

8. The **adductor** lies immediately cranial to the semimembranosus, mostly between it and the femur. This muscle does not really belong to the superficial group; however, it is closely associated with the previously named muscles. On its lateral surface its fibers usually extend obliquely to the semimembranosus, whereas on

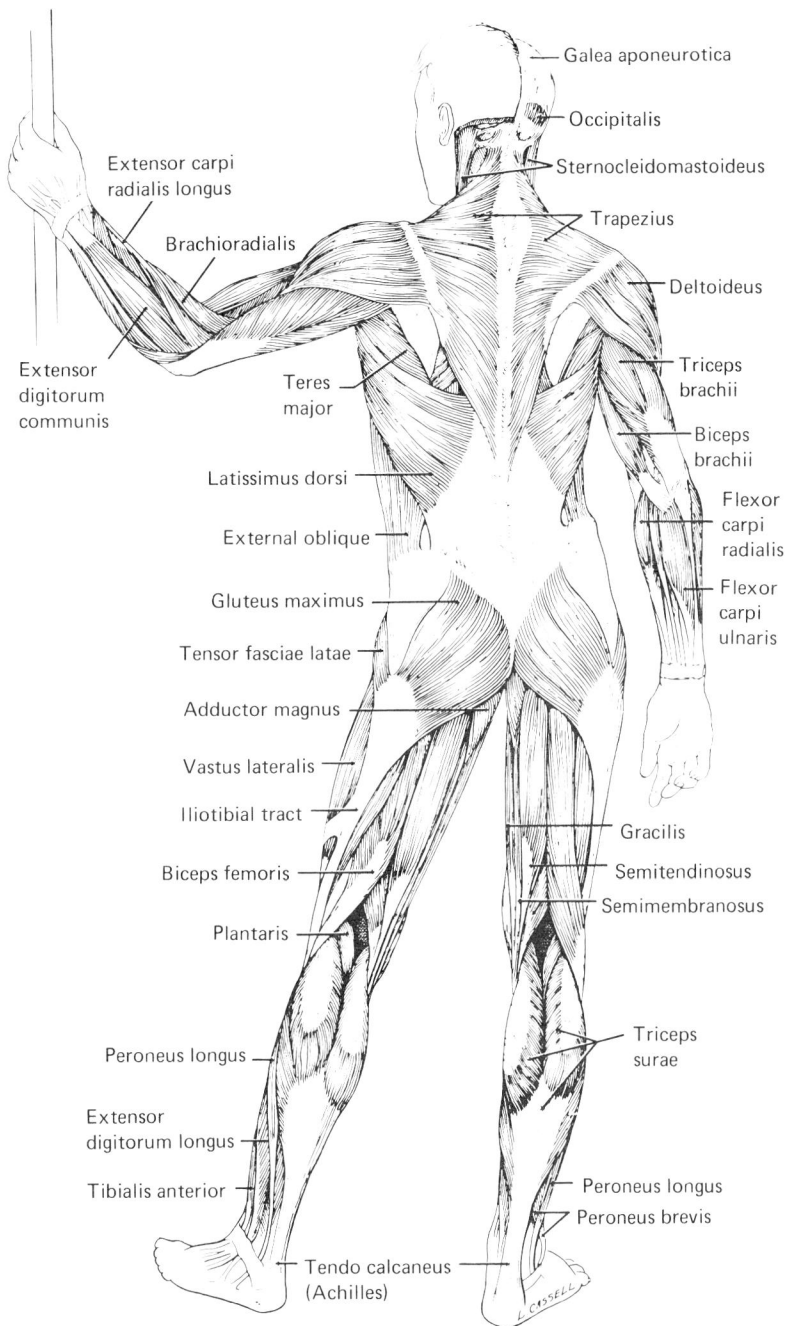

Fig. 2-13. Muscles of the human body, posterior view. (From Millard, N. D., King, B. G., and Showers, M. J.: Human anatomy and physiology, Philadelphia, 1956, W. B. Saunders Co.)

its medial surface they are almost parallel to it. The adductor arises on the ischium and pubic symphysis and passes downward and caudally to be inserted along most of the entire length of the femur. It is powerful, although its fibers are not closely bound together. It is an adductor of the thigh.

Thigh muscles of a beef animal constitute "round steaks."

DEEP MUSCLES OF HIP AND THIGH (Fig. 2-14)

1. The **gluteus superficialis** (**maximus**) was first named on man, where it is the largest of

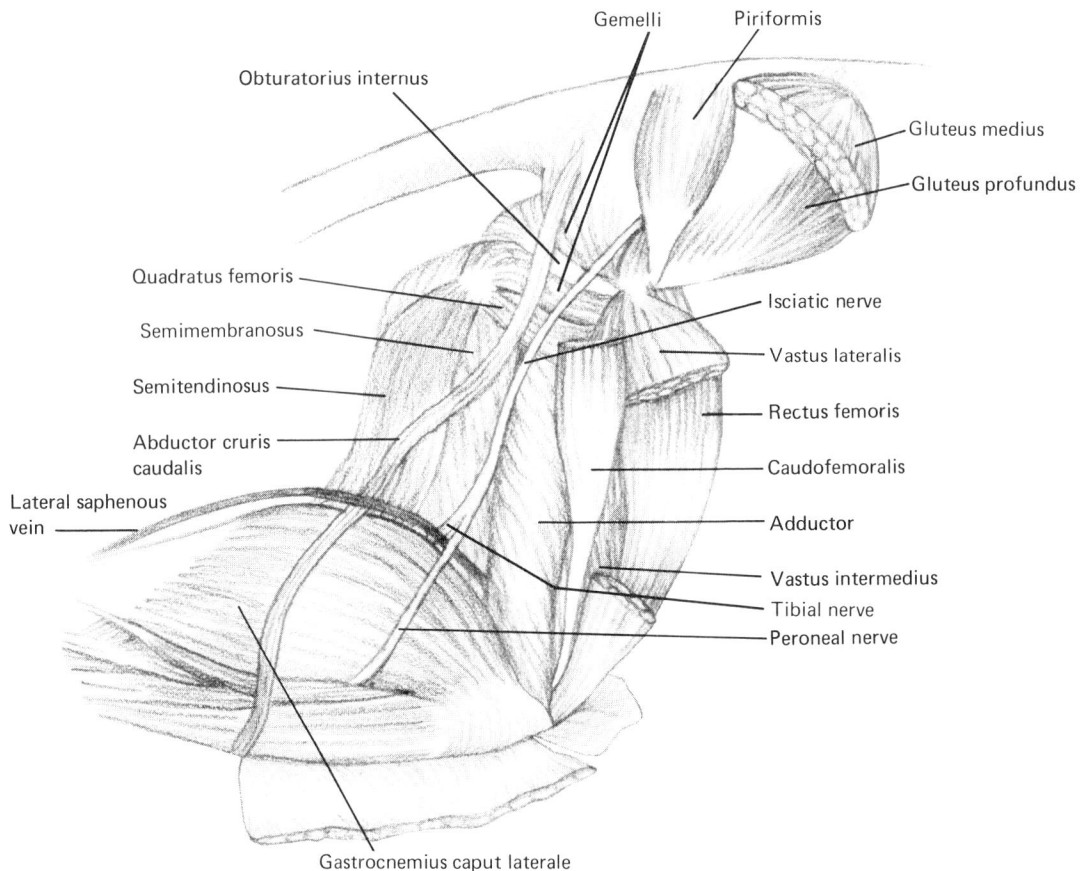

Fig. 2-14. Deep muscles of hip and thigh, lateral view.

the three gluteal muscles and forms the principal cushion when he sits (see Figs. 2-12 and 2-13), but in the cat the **gluteus medius** is the largest. The gluteus superficialis arises from the sacrum and the first two caudal vertebrae and extends laterally over the pelvis and inserts below the greater trochanter of the femur. It is about one-half inch wide and two inches long in a medium-sized cat. It is an abductor of the hip. Loosen its edges, bisect, and reflect. This will expose more of the **isciatic nerve.**

2. The **gluteus medius** is the largest of the gluteal muscles in the cat. It arises on the cranial end of the ilium and passes caudally and laterally dorsal to the ilium and inserts on the greater trochanter of the femur. It is twice the size of the gluteus superficialis. When the upper end of the tensor fasciae latae was reflected, the lateral edge of the gluteus medius was exposed. Dissect the fascia covering from the gluteus medius. Lift up the thick lateral edge from the gluteus profundus, which usually has a white sheath, and insert a probe between them close

to the femur, which should be pushed up in order to make this procedure easier. Do not mutilate the underlying muscles. It is often difficult to bisect and reflect the gluteus medius because of its thickness and the presence of the erector spinae muscle in the sacral region. If, after bisecting, the portion close to the femur separates easily into two parts and the lower part has many white fibers in contact with the isciatic nerve, then it is quite certain that the piriformis has been bisected also.

3. The **piriformis** lies deeper than the gluteus superficialis and gluteus medius. It may have been bisected with the latter because the lower portions are almost merged with one another. If the piriformis has been bisected and reflected with the gluteus medius, the base of the isciatic nerve can be traced over the ilium between it and the sacrum. The piriformis may be identified by the white, shiny tendon that joins the femur and by its triangular shape. It arises from the ventrolateral surface of the sacrum and the first few caudal vertebrae, passes dorsal to the

isciatic nerve, and inserts on the greater trochanter of the femur. Bisect it if that has not been done. It helps to draw and hold the head of the femur in the acetabulum. It is much the same in man.

4. The **gluteus profundus (minimus)** is under the lateral edge of the gluteus medius. It is cylindrical and almost cigar shaped. It arises on the lateral and dorsal surfaces of the ilium and inserts on the greater trochanter of the femur. It is much the same in man, since it helps to rotate the femur. **Do not bisect** the gluteus profundus.

5. The **obturatorius internus** may be identified as follows. Bisect the isciatic nerve near the head of the femur and pull the proximal portion up against the reflected piriformis and gluteus medius. Cut away the fascia and fat dorsal and medial to the caudal end of the ischium and push the base of the tail away from the ischium. The obturatorius internus is horeshoe shaped, as shown in Fig. 2-14, and passes over the caudal inner surface of the ischium. The origin is on the lower medial portion of the ischium close to the pubis. It passes dorsalward between the base of the tail and ischium and curves over the ischium and under the ischiatic nerve to insert on the greater trochanter of the femur. It is usually about one-half inch wide and helps to pull the head of the femur medially and caudally. Loosen the dorsal half but do not bisect.

6. The **gemellus caudalis (inferior)** lies cranial to the dorsal half of the obturatorius internus, and the two unite in a common flat, white tendon and insert on the femur. The origin is on the second or third caudal vertebra from which the muscle passes over the pelvis, where some of the fibers also arise, and continues laterally to merge with the insertion of the obturatorius internus. It is about one centimeter wide and is slightly constricted as it passes over the pelvis. Loosen the edges but **do not bisect.** If possible, obtain some modeling clay and model these small muscles on the mounted cat's skeleton to get a definite idea of their relationships.

7. The **gemellus cranialis (superior)** lies immediately cranial to the gemellus caudalis (inferior). Its lateral surface is against the inner side of the gluteus profundus and is below the base of the isciatic nerve. It arises on the crest of the ilium and inserts on the head of the femur.

Separate its edges from the previously named muscles but **do not bisect.** These smaller muscles of the hip help to hold the head of the femur in its socket, the acetabulum.

8. The **quadratus femoris** lies immediately dorsal and lateral to the adductor and lateral to the inner end of the obturatorius internus. Bisect the adductor in a plane extending from the center of the ventral edge dorsalward to the insertion of its upper edge near the head of the femur. Pull the femur forward and reflect both ends of the adductor. The lateral surface of the quadratus is now exposed. This muscle is four sided, almost a cube, in relation to its origin and insertion, which are on the ischium and greater trochanter, respectively. It is an extensor and rotator of the thigh. It is the same in man. **Do not bisect.**

9. The **obturatorius externus** lies below the quadratus femoris and ischium, but medial to the caudal portion of the adductor. Pull the caudal portion of the adductor ventralward to expose the lateral triangular surface of the obturatorius externus. It arises from the lower caudal part of the ischium and the pubis near the symphysis. Its fibers extend dorsolaterally and forward to insert on the femur lateral and distal to the lesser trochanter. **Do not bisect.** Pull the femur forward and outward to see the triangular caudal surface of the adductor longus (Fig. 2-15) and the femoral artery and vein passing almost parallel to its lower edge. The branches of the isciatic nerve are shown.

SUPERFICIAL AND DEEPER MUSCLES OF THIGH (Figs. 2-15 and 2-16)

1. The **sartorius** has been considered in a previous group (see p. 67) because it appears on the lateral surface and on the cranial half of the medial surface. It has been bisected, but in Fig. 2-15 it is shown intact. In man it is the longest muscle in the body, since it arises on the outer anterior crest of the ilium and passes obliquely across the front of the thigh to the medial side of the knee and attaches by an aponeurosis to the tibia. This can be seen in Fig. 2-4, the anterior view of man's muscles.

Cranial to the femur and partially covered by the sartorius, as shown in Fig. 2-16, is the **vastus medialis.** It is close against the femur and arises from it and the ilium. The **rectus femoris** (Fig. 2-14) is cranial and lateral to the vastus medialis and arises entirely from the ilium. The circum-

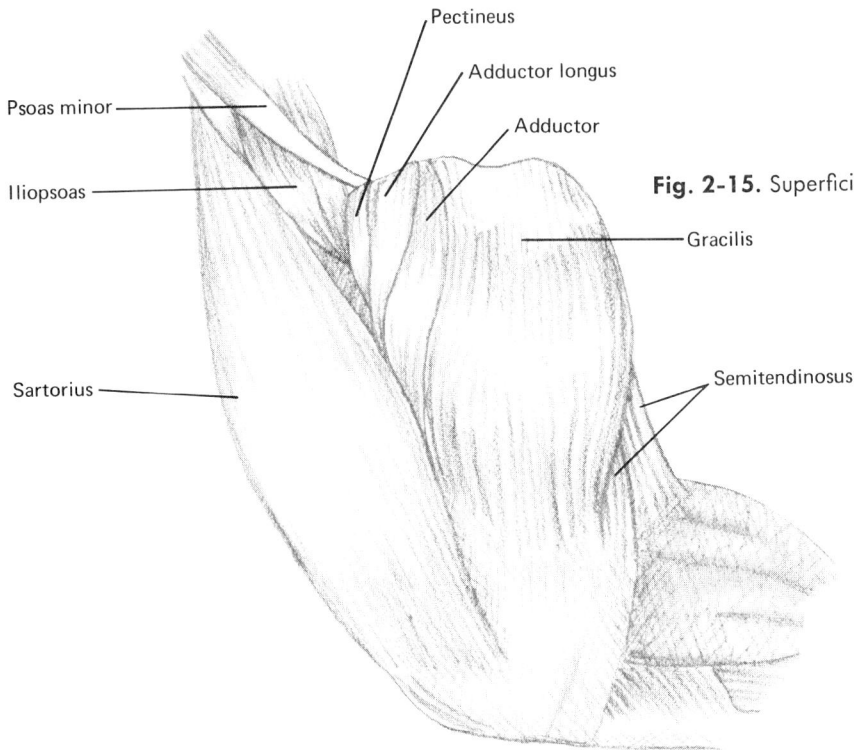

Fig. 2-15. Superficial muscles of thigh, medial view.

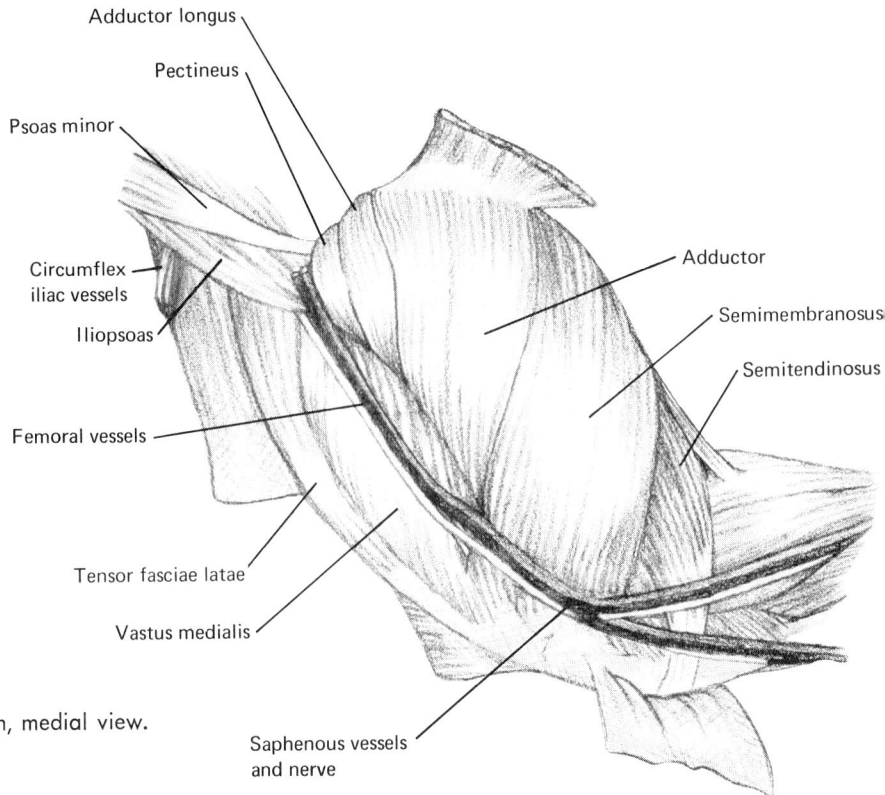

Pectineus

Adductor longus

Adductor

Psoas minor

Iliopsoas

Gracilis

Sartorius

Semitendinosus

Adductor longus

Pectineus

Psoas minor

Circumflex
iliac vessels

Iliopsoas

Femoral vessels

Adductor

Semimembranosus

Semitendinosus

Tensor fasciae latae

Vastus medialis

Fig. 2-16. Deep muscles of thigh, medial view.

Saphenous vessels
and nerve

flex artery passes between these two muscles. Still farther forward and arising from the ilium is the **tensor fasciae latae**, of which only a portion is shown in Figs. 2-12 and 2-16.

2. The **gracilis** was bisected in order to see the medial sides of the **semitendinosus** and **semimembranosus**; hence it is shown reflected in Fig. 2-14. It is broad and thin, covering the caudal portion of the inner or medial surface of the thigh. In several respects it seems comparable with the tensor fasciae antebrachii. The gracilis arises near the symphysis pubis and inserts on the inner side of the proximal side of the tibia. Part of the fascia joins with that of the tensor fasciae latae. It is an adductor of the thigh. It is the same in man.

3. The **adductor**, previously mentioned, is a large muscle and arises from the rami of the pubis and ischium and inserts on the medial inner surface of the femur. It lies along the cranial edge of the semimembranosus and inserts on about the middle third of the shaft of the femur. It corresponds with the adductor magnus and adductor brevis of man. Its action is to extend the thigh after it has been drawn forward. The cranial edge of the adductor may separate off and appear as an additional muscle.

4. The **adductor longus** is much smaller than the adductor and is only about one inch long. It lies in contact with the cranial proximal surface of the adductor. It extends from the pubic bone to the distal end of the upper third of the femur. **Do not bisect.** This small muscle is a part of the adductor in some animals.

5. The **pectineus** is still smaller than the adductor longus and lies cranial to it and under the cranial proximal edge of the gracilis. It arises on the pubis and inserts on the upper end of the femur.

6. The **iliopsoas** is a long and cylindrical muscle and lies along the inner surface of the dorsal body wall in the lumbar region. It arises from the undersides of the transverse processes of the last two thoracic and the lumbar vertebrae. It is a hypaxial muscle, being below the transverse processes, in distinction to the erector spinae group, which is epaxial, or above, the transverse processes. The iliopsoas passes caudally through the body wall dorsal to the inguinal region but ventral to the ilium. It may be partially uncovered as it inserts on the lesser trochanter of the femur. **Do not bisect.** In man these muscles separate into psoas major and iliacus. A long muscle with a long tendon of insertion lies medial to the iliopsoas and inserts at the side of the pelvic inlet. This is the **psoas minor.**

FEMORAL TRIANGLE OF THIGH

The femoral triangle is an important area on the medial surface of the thigh. It is bounded by the proximal half of the caudal edge of the sartorius and a line along the femur where the adductor and adductor longus attach to the femur. The third side of the triangle is the lateral edge of the ilium. Extending across this triangle are the following structures: (a) The **femoral (external iliac) artery** is usually injected with red latex. It gives off the circumflex to the deep muscles that arise on the ilium and upper part of the femur. (b) The **femoral (external iliac) vein** is usually larger than the artery and dark because of the blue injecting material or the coagulated blood. It receives the **deep femoral vein** from under the proximal end of the gracilis and adductor muscles. This vein is parallel with the femoral artery. At about the middle of the thigh one or more **muscular veins** enter the femoral from the adductor and semimembranosus muscles. (c) The **femoral nerve** is usually cranial or dorsal to the blood vessels at its base and farther out gives off branches to the superficial muscles of the thigh and shank. More details will be given on these blood vessels and nerves later.

DEEP MUSCLES OF CRANIAL PORTION OF THIGH

The **quadriceps femoris** (quadriceps extensor group) consists of four muscles that have separate origins but insert on the tibia by the common tendon of the patella.

1. The **vastus lateralis** (Fig. 2-14) is partially under the tensor fasciae latae and is caudal to the sartorius. It is the largest of the group and the most external of the four. It arises on the outer surface of the shaft of the femur and on the greater trochanter. Passing down, it joins with other muscles to insert on the lateral surface of the patella in the so-called patella tendon. It is the same in man. It acts as an extensor of the knee. Separate it from the two underlying muscles, beginning at its upper anterior surface. In addition, cut it loose along its caudal edge from the femur. Bisect and reflect.

2. The **rectus femoris** (Fig. 2-14) is the most cranial of the four muscles, is uniform in size,

and lies between the vastus lateralis and sartorius. It arises from the **ilium** above the acetabulum. It passes caudally to join the vastus lateralis and inserts in common with it in the patella tendon. It is the same in man. It acts as an extensor of the knee and a flexor of the hip. Bisect.

3. The **vastus medialis** (Fig. 2-16) lies caudal and medial to the rectus femoris and is covered medially by the sartorius. The sheath or fascia of the tensor fasciae latae is attached to it. Its origin is from the inner and cranial surface of the proximal end of the femur. It inserts on the medial margin of the patella tendon. It is the same in man. It acts as an extensor of the knee. **Do not bisect.**

4. The **vastus intermedius** (Fig. 2-14) is the smallest of the four muscles. It arises along the cranial and lateral surfaces of the femur, to which it adheres closely. It is medial to the vastus lateralis and soon merges with the vastus medialis, and the two often appear as one muscle. To expose this muscle, reflect muscles one, two, and three. It is the same in man. It acts as an extensor of the knee. **Do not bisect.**

5. The **articularis coxae** (not shown) is closely associated with the quadriceps femoris group. It arises on the outer caudal surface of the ilium and passes caudally between the origins of the rectus femoris and the vastus lateralis to insert on the lateral surface of the femur above the vastus intermedius. It is small, only about one inch long, cranial to and below the head of the femur. It helps to rotate the thigh and pull it forward.

The muscles of the thigh of a beef animal constitute the **round steak** at the meat market.

Name _____

Date _____

REVIEW QUESTIONS ON MUSCLES OF HIP AND THIGH

1. Name five superficial muscles of the hip and thigh of the cat and of man.

2. Name five small, deep muscles of the hip.

3. Describe the structure of the tensor fasciae latae muscle.

4. What long slender muscles lie medial to the biceps femoris muscle?

5. What are the relationships of the caudofemoralis muscle to the tensor fasciae latae and biceps femoris?

6. What are the names of the two large muscles on the medial surface of the thigh?

7. State the principal differences of the glutei maximus (superficialis) muscles of cat and man.

8. Compare the sartorius muscles of the cat and man.

9. Name the small muscles that are inserted on the proximal end of the femur.

10. Name the three gluteal muscles.

11. Which muscles are known as the "hamstring" muscles?

12. What muscle curves over the inner and outer surfaces of the ischium?

13. What are the two main divisions of the isciatic nerve? (Fig. 2-14)

14. Where is the femoral triangle and what are its boundaries?

15. What ventral thigh muscles originate on the pubic bone?

16. What is the advantage of having heavy fascia (aponeurosis) on the outside rather than on the inside of the thigh? (Use your judgment.)

17. Heavy fascia, or aponeurosis, of the thigh represents a modified part of what muscle?

18. From what part of a beef animal are round steaks cut?

19. Why do the names of the gluteal muscles seem inappropriate?

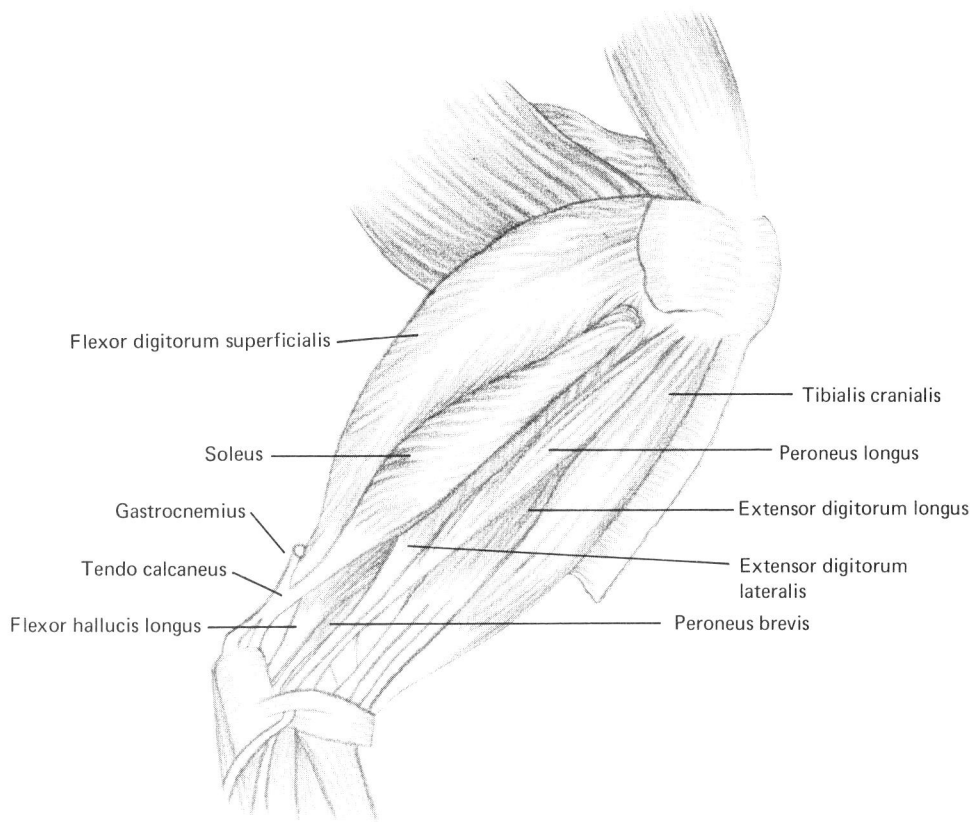

Fig. 2-17. Muscles of leg, lateral view.

MUSCLES OF LEG (Fig. 2-17)

Reflect the lower portion of the biceps femoris, including its broad sheath on the outer surface of the leg, and the semitendinosus and semimembranosus completely in order to see the following structures. The **lateral saphenous vein,** mentioned in connection with the thigh, comes up the caudal surface of the shank to join the **popliteal vein** behind the knee joint. A **communicating branch** of the **lateral saphenous vein** extends up the caudal surface of the thigh and connects with the **caudal gluteal vein** of the hip. Often these veins are not injected or dilated with coagulated blood and hence are not easily seen.

1. The **flexor digitorum superficialis (plantaris muscle)** arises from under the cranial edge of the biceps on the patella and passes across the lateral side of the knee and down between the two heads, medial and lateral, of the gastrocnemius, merging with them. These help form the **tendo calcaneus (tendon of Achilles)** and insert on the **calcaneus,** or heel bone, of the

hock. Loosen the edges of the plantaris lateral to the knee until it merges with the gastrocnemius. The **tibial branch of the isciatic nerve** passes between the plantaris and the medial head of the gastrocnemius. Separate the plantaris from the medial head and bisect the lateral head where the two merge with one another. The **femoral artery** and **vein** pass near the femur.

2. The **gastrocnemius** is the largest muscle of the leg. It has two heads, called the caput mediale and the caput laterale, one on either side of the plantaris. These two heads merge with the plantaris, usually throughout their lower two thirds. The **peroneal branch of the isciatic nerve** passes lateral to the lateral head of the gastrocnemius and penetrates its craniolateral edge.

3. The **soleus** arises cranial to the lateral head of the gastrocnemius from the lateral surface of the head of the fibula. It extends medial to the lateral head of the gastrocnemius, and one third or more of the muscle is exposed laterally

before it joins the lateral surface of the gastrocnemius tendon to form the tendon of Achilles. The plantaris, gastrocnemius, and the soleus form most of the **calf of the leg.** The two heads of the gastrocnemius and the soleus constitute the **triceps surae.**

DEEP CRANIAL MUSCLES OF LEG (Fig. 2-17)

Remove two or three layers of the heavy fascia covering the lateral and cranial surfaces of the leg. Consider the lateral view of the leg and proceed from cranial to caudal.

1. The **tibialis cranialis** may be identified as follows. The sharp cranial edge of the tibia is often called the shinbone. Locate this sharp-edged bone on the cat and on yourself. This muscle lies immediately lateral on the tibia and occupies about three fourths of the space between the tibia and fibula. The origin is on the upper craniolateral surface of the tibia and the shaft and head of the fibula. It passes distally down the craniolateral surface of the tibia and inserts on the outer surface of the first metatarsal. Separate the tibialis cranialis from the extensor digitorum longus, which lies caudal to it. Insert a probe under the tibialis and bisect below the center. It is triangular in cross section and usually is much heavier near the proximal end.

2. The **extensor digitorum longus** lies deep to the tibialis cranialis. Its proximal end is almost completely encased by the tibialis cranialis. Medially and distally from the center of the leg its lateral surface is exposed to the superficial fascia, whereas its cranial and medial surfaces are still bordered by the tibialis cranialis. It originates in the extensor fossa of the femur and extends along the cranial side of the fibula. With the tibialis cranialis it passes under the **annular,** or **transverse, ligament** on the outside of the hock. **Do not bisect.** Compare Fig. 2-17 with Fig. 2-13.

3. The **peroneus longus** is a slender, uniformly shaped, superficial muscle caudal to the extensor digitorum longus. It originates on the head and lateral surface of the fibula, and it inserts on the metatarsals. It acts as a flexor of the tarsus, as in man. The inner surface lies above the two following muscles. **Do not bisect.**

4. The **extensor digitorum lateralis** is a long, slender muscle lying directly beneath the peroneus longus and is usually smaller. It originates

by fleshy fibers from the second quarter of the lateral surface of the fibula and continues as a thin, shiny tendon that passes through the groove of the lateral malleolus of the fibula. The proximal end is covered by the peroneus longus and distally is covered by superficial fascia and lies caudal to the peroneus longus.

5. The **peroneus brevis** is a short, thick muscle lying beneath the other lateral extensors, originating by fleshy fibers from the distal half of the fibula under the upper end of the soleus. It adheres to and extends along the lateral part of the fibula and ends by passing through the groove on the ventral border of the lateral malleolus with the lateral extensor. The lateral surface is covered by the peroneus longus, extensor digitorum lateralis, and superficial fascia; the inner surface lies against the fibula. **Do not loosen from the fibula or bisect.**

DEEP CAUDAL MUSCLES OF LEG (Fig. 2-18)

Reflect the gastrocnemius, superficial digital flexor (plantaris), and soleus muscles. The following muscles will be considered in order, beginning on the upper medial surface and proceeding laterally.

1. The **popliteus** is a short, broad, tapering, triangular muscle arising on the lateral epicondyle of the femur almost beneath the lateral head of the gastrocnemius. It extends over the lateral articular facet on the proximal end of the tibia and inserts on the medial, proximal, and upper end of the medial portion of the tibia. The lateral surface is medial to the gastrocnemius and plantaris and the insertion of the semitendinosus. Its action rotates the leg.

2. The **flexor digitorum longus** is a long, tapering muscle lying lateral to the distal edge of the popliteus and extending from its origin on the lateral surface of the fibula to a fine tendon. It originates on the proximal, ventral surface of the tibia, extends down the medial surface of the tibia, receiving fibers therefrom, and ends in a thin, shiny tendon that passes over the ventral groove on the distal end of the fibula to join the tendon of the flexor hallucis longus. It is bordered medially by the medial head of the gastrocnemius and superficial fascia.

3. The **tibialis caudalis** is a long, thin, flat muscle beneath and lateral to the flexor digitorum longus and between it and the flexor hallucis longus. It originates on the medial sur-

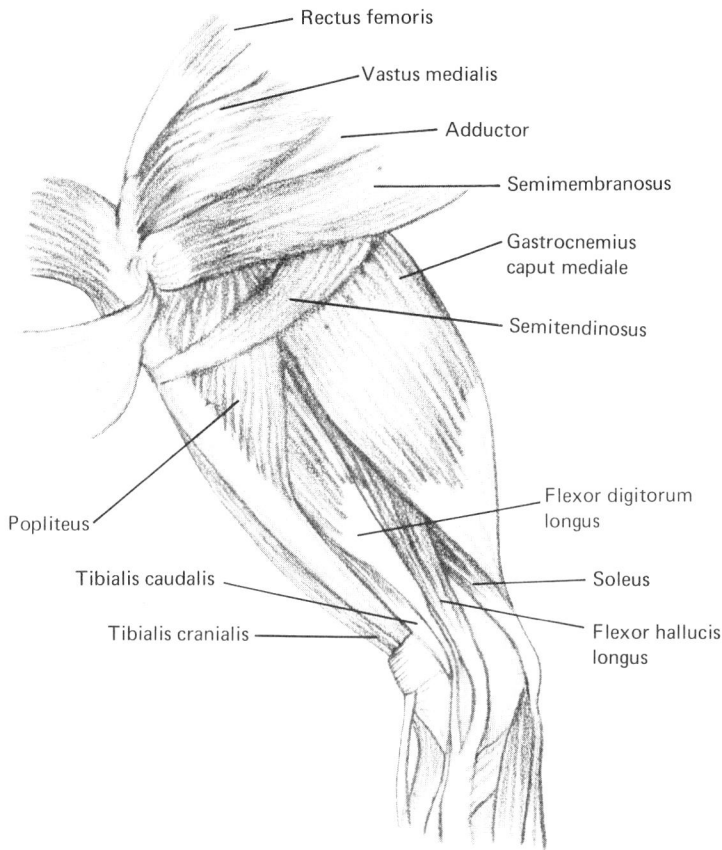

Fig. 2-18. Muscles of leg, caudomedial view.

face of the head of the fibula and ventral surface of the tibia, extending as a thin, flat muscle to the middle of the fibula, where it merges with a thin, flat, shiny tendon and inserts on the tarsals.

4. The **flexor hallucis longus** is a long, heavy muscle lateral to the tibialis caudalis and connected to the caudal border of the tibia by the fleshy fibers from the cranial end to within one to three centimeters of the distal end. It is also attached to the proximal caudal end of the fibula by fleshy fibers. It passes down the caudal surface of the leg to the tarsus as a uniform, broad muscle cranial to the tibial nerve. It continues through the tarsal canal as a tendon that goes with the tendon of the plantaris to the digits. The flexor digitorum longus, flexor hallucis longus, and, in some animals, the tibialis caudalis are grouped as **flexor digitorum profundus.**

The **tibial nerve** passes between the medial and lateral heads of the gastrocnemius and continues along the flexor digitorum longus, sending branches to the leg and foot. The **popliteal artery**, which is a continuation of the femoral, passes through the popliteal space and then passes deep to the popliteal and flexor digitorum longus muscles. It continues between the tibia and fibula, where it is known as the cranial tibial artery. The **popliteal vein** lies close to the popliteal artery and receives the lateral saphenous and the caudal femoral veins.

REVIEW QUESTIONS ON MUSCLES OF LEG

1. What muscles constitute the calf of the leg?

2. Describe the general form of the gastrocnemius muscle. Name its two parts.

3. What muscles contribute to the tendo calcaneus?

4. Name three kinds of connective tissue associated with muscles.

5. What nerves supply most of the muscles of the calf of the leg? (Fig. 2-14)

6. What is the relation of the tendo calcaneus to the heel of man?

7. What is the main function of the muscles of the calf of the leg. (Use your own judgment to answer.)

8. What artery and vein pass down in back of the knee?

9. What is an aponeurosis and where may it be found in the cat?

10. What bones support the leg (shank) muscles? (Figs. 1-1 and 1-2)

11. What bone is intermediate between the femur and the bones of the leg? (Figs. 1-1 and 1-2)

12. What is another name for the shinbone?

13. What muscle lies between the two parts of the gastrocnemius and merges with them?

14. Do you think there is a difference in the demand on muscles in an animal whose body is horizontal and one in an erect position? (Use your own judgment to answer.)

SOME DIFFERENCES IN MUSCLES OF THE CAT AND MAN

Many muscles of man have apparently become modified largely because of the erect position of his body.

1. The cleidocervicalis, acromiotrapezius, and spinotrapezius muscles in the cat are merged into one muscle in man.

2. The cleidobrachialis, continuous with the cleidocervicalis on the upper arm of the cat, is absent in man.

3. The acromiodeltoideus and spinodeltoideus muscles of the cat are merged into one deltoid muscle in man.

4. The tensor fasciae antebrachii, abductor cruris caudalis, cutaneus trunci, rhomboideus capitis, and omotransversarius muscles present in the cat are absent in man.

5. The sternomastoideus and cleidomastoideus are separate in the cat, whereas in man their upper ends have merged, hence the name sterno-cleidomastoideus.

6. In the cat the superficial and deep pectoral muscles are somewhat subdivided. In man they are more closely associated into the pectoralis major and minor.

7. Two rhomboideus muscles, the major and minor, are found in cat and man. The muscles were named first on man. The major is the larger in man, whereas in the cat the minor is the larger. Hence, the terms have been changed to thoracis (major) and cervicis (minor).

8. The sartorius in man extends obliquely across the inner surface of the thigh from the crest of the ileum to the inner region of the knee, whereas in the cat it extends uniformly along the lateral and cranial surfaces of the thigh.

9. Man has an omohyoideus muscle arising on the scapula, extending diagonally across the neck behind the sternohyoideus, and inserting on the hyoid bone. The cat has no omohyoideus muscle.

10. The gluteus maximus muscle is the largest of the glutei in man, but in the cat the gluteus maximus, called superficialis, is smaller than the gluteus medius.

11. The biceps femoris in the cat is relatively large, whereas in man it is relatively small.

12. The serratus posticus superior and inferior (serratus dorsalis, cranialis and caudalis) are adjacent to one another in the cat, whereas in man they are separated.

13. The rectus abdominis in man extends from the fourth rib to the pubic symphysis, whereas in the cat it extends from the third rib to the symphysis. This muscle is definitely divided by four transverse tendinous inter-sections in man, whereas in the cat these lines are present but not as easy to identify.

14. The flexor digitorum profundus has five parts in the cat, whereas in man it is considered as one muscle, which ends in four tendons on the digits.

15. The medial head of the triceps in man is not divided.

16. The flexor carpi ulnaris, which appears almost as two muscles in the cat, is not so completely separated in man.

17. The adductor of the cat corresponds to the adductor magnus and the adductor brevis of man.

GENERAL CONCLUSIONS ON MUSCLES

1. It is commonly accepted among anatomists that man's general structure of musculature has become modified, largely because of a gradually acquired erect position and because muscles have gradually changed to better meet the demands of activities throughout man's phylo-genetic development.

2. A reduction in the number of parts, such as in the muscles of man when compared with those of the cat, is interpreted as an indication of advancement, or higher specialization.

THREE

General internal organs of the cat

SURVEY OF INTERNAL ORGANS
(Figs. 3-1 to 3-3)

In the study of blood vessels and internal organs, the inexperienced student is often confused about which is the right or the left side of the specimen, particularly when it is lying on its back. It is helpful in such cases to think of your own body as lying in the same position as the specimen you are dissecting or that is being described. You can always state which is your right or left, dorsal or ventral.

Place the injected cat on its back on the dissecting tray and gradually separate the hind limbs. It is best to tie a stout string about three feet long to one of the hind limbs and to pass the string under the tray and tie the other end of the string to the other hind limb. Feel the cranial end of the **pubic symphysis** with your finger. Procure a heavy knife and separate the two pubic bones of this symphysis by cutting between them. This is often difficult to do. If there is a mounted cat skeleton for you to examine, look at the **pelvic symphysis** carefully before trying to do this. The laboratory assistant can give some additional advice to help, since it is important to do this correctly.

After the symphysis has been separated, open the body cavity by cutting through the body wall along the midventral line, or raphe, from the symphysis to the sternum, being careful not to cut the internal organs. Continue the incision along the left side of the sternum, cutting through the costal cartilages. As you do this, you will see the **internal thoracic artery** and **vein** along the inner side of the sternum. The **ventral mesentery** will also be seen as a thin membrane extending from the ventral body wall to within the median cleft of the liver, where it is known as the **falciform ligament**. The muscular **diaphragm** consists of skeletal muscle and tendon

and separates the thoracic and abdominal cavities. Cut the diaphragm on both sides about one inch from the body wall and observe the angle at which it is attached to the body wall. Is its ventral, lateral, or dorsal attachment most cranial? What are the relationships of these attachments to the ribs? Cut on the other side of the entire sternum and remove it.

Both sides of the thoracic wall are to be reflected so that the internal organs may be more easily examined. The ribs will usually break as this is done. The **serratus ventralis** muscle holds the limb onto the body; if this muscle is bisected, the limb is held by vessels and nerves only.

The fatty mass covering the **intestine** is the **greater omentum**. The thin, shiny parietal peritoneum lines the thoracic and abdominal body walls, including both sides of the diaphragm. It is reflected over each of the internal organs, where it is known as serous membrane, or visceral peritoneum. Ventral to where the esophagus passes through the diaphragm there is a clear, transparent area known as the **central tendon of the diaphragm**. The diaphragm is covered on the thoracic side by pleura and on the abdominal side by peritoneum. The tendinous area is the last to close in the embryonic development of the diaphragm, and sometimes this opening persists even in man and is known as a **patent diaphragm**. Adult cats that were apparently otherwise normal have been found with this defect. Sometimes the abdominal organs invade the pleural cavity through an opening made in the diaphragm due to injury. This condition is called a **diaphragmatic hernia**. Any marked deviation in structure from the normal is known as an anomaly.

The liver is attached at the periphery of its base to the diaphragm by a suspensory ligament, the **coronary ligament**, which is a continuation

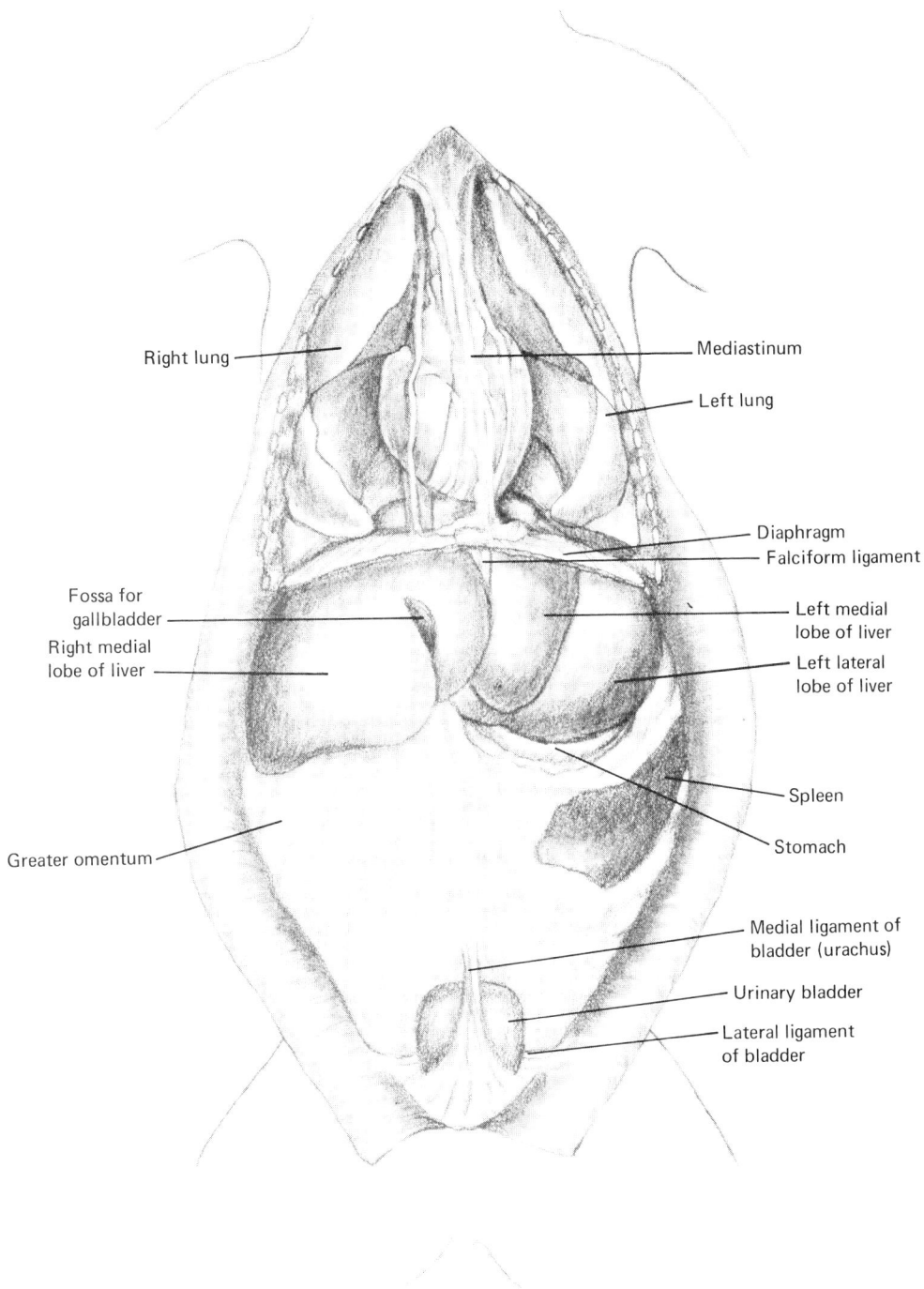

Fig. 3-1. Thoracic and abdominal viscera, organs intact.

of the falciform ligament, and is the reflection of the peritoneum of the abdominal side of the diaphragm onto the surface of the liver. This attachment is largely a result of its growth in the embryological development of the liver, from the duodenum into the ventral mesentery, a part of which is the septum transversum and which in turn becomes part of the diaphragm.

The **common hepatic duct,** the **gallbladder,** and the **liver** tissue are largely modified parts of the wall of the duodenum and the ventral mesentery from which they develop embryologically.

The various lobes of the liver may differ in size from that shown in Fig. 3-1. Identify the lobes of the liver as follows: **left medial,** lying along the falciform ligament and against the left portion of the diaphragm; **left lateral lobe,** be-

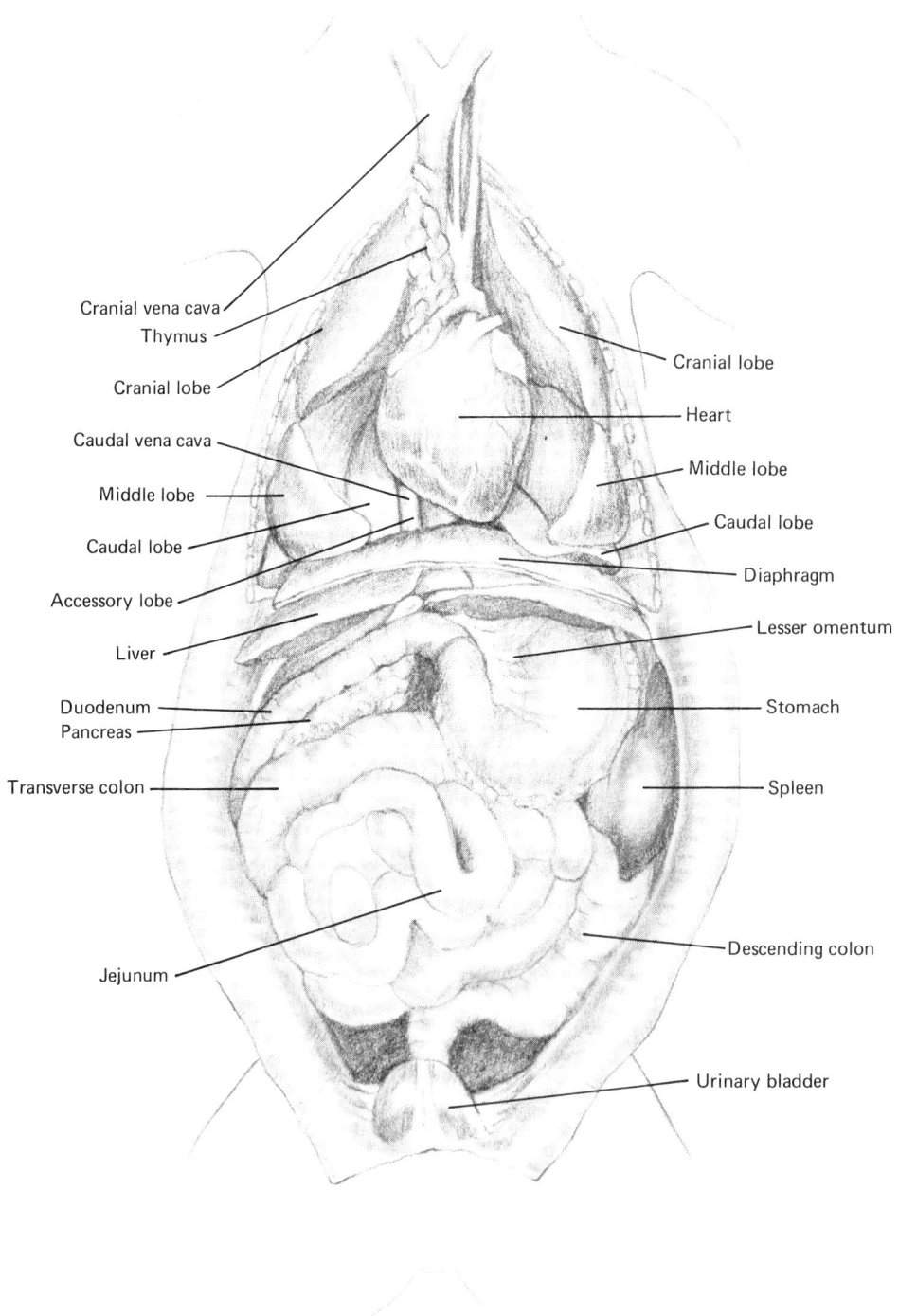

Fig. 3-2. Thoracic and abdominal viscera, superficial structures removed.

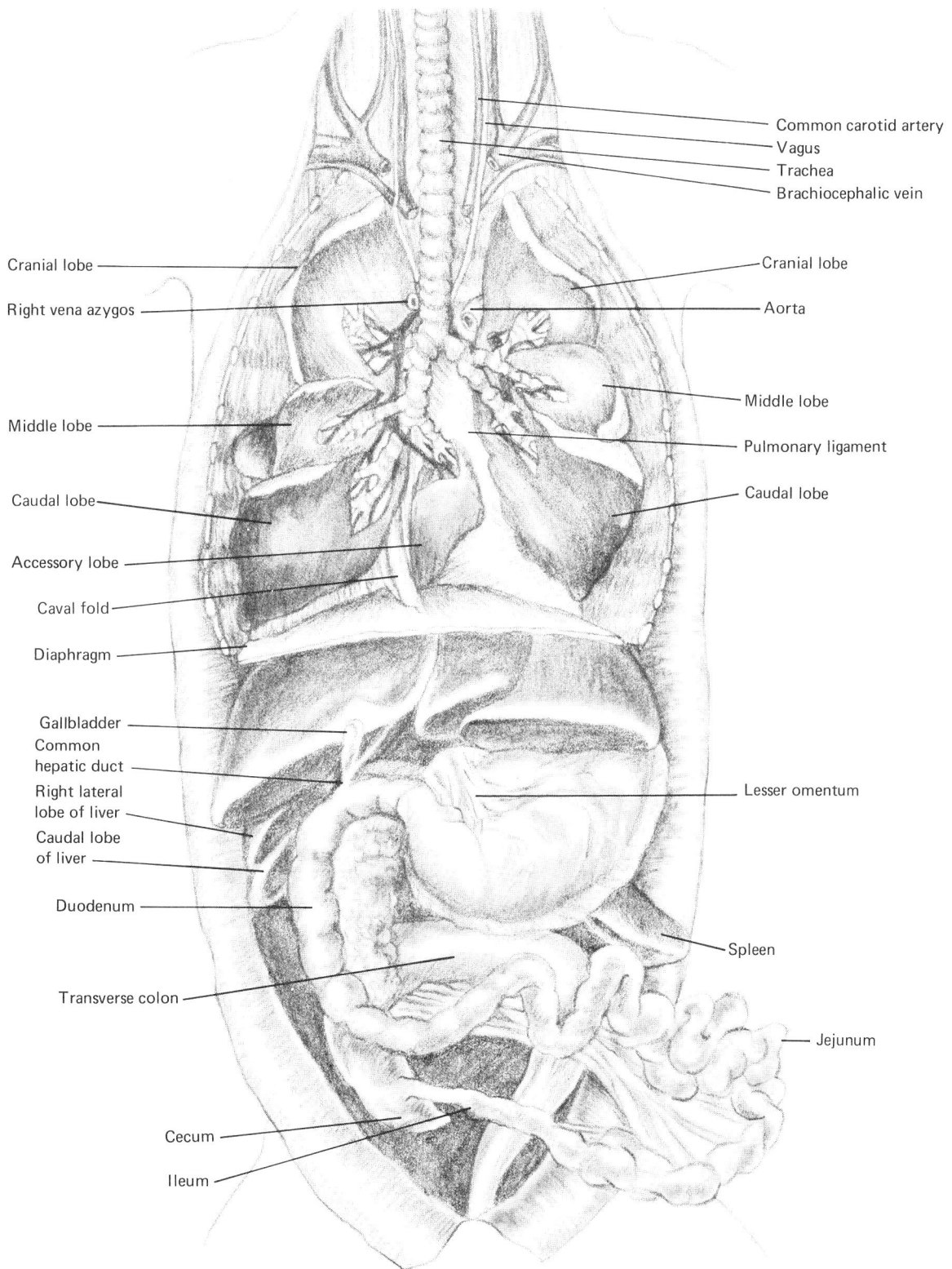

Common carotid artery
Vagus
Trachea
Brachiocephalic vein

Cranial lobe

Right vena azygos

Cranial lobe

Aorta

Middle lobe

Middle lobe

Pulmonary ligament

Caudal lobe

Caudal lobe

Accessory lobe

Caval fold

Diaphragm

Gallbladder
Common
hepatic duct

Lesser omentum

Right lateral
lobe of liver

Caudal lobe
of liver

Duodenum

Spleen

Transverse colon

Jejunum

Cecum

Ileum

Fig. 3-3. Thoracic and abdominal viscera, deep dissection.

tween the diaphragm and the cardiac end of the stomach; **right medial (cystic) lobe,** lying against the right half of the diaphragm and containing the **gallbladder** within its cleft. The smaller **right lateral lobe** is caudal and lateral to the right medial lobe. Its caudal position is partially marked off by a constriction, which extends backward to the right kidney. The **caudal lobe** is the smallest of the five lobes. It has a portion projecting toward the right kidney and one (papillary) into the lesser omentum.

Find the esophagus as it pierces the diaphragm and enters the stomach. Determine the stomach as to position, size, and shape. The end next to the esophagus is called the **cardiac extremity;** the opposite end is the **pyloric.** The **fundus** is the dilated portion of the stomach. The lesser curvature of the stomach is the embryonic **ventral margin,** whereas the greater curvature indicates the original embryonic **dorsal margin** where the ventral part of the greater omentum is attached. The position changes in development.

Cats with long fur often swallow enough fur when licking themselves to form "fur balls" in the stomach; they may not be passed through the intestine but may cause sickness and are eventually vomited.

The **greater omentum** is really a part of the two-layered **dorsal mesentery** and extends ventrally from the spinal column. It becomes folded and attached to the greater curvature of the stomach. The **spleen** lies between the two layers of the greater omentum as it extends down as a double-walled bag between the intestines and the ventral body wall (Fig. 3-1). The greater omentum assists in regulating the temperature of the body, and it is also protective because of the many phagocytic cells contained in it. Carefully lift up and unfold the **greater omentum,** beginning at its caudal edge, and you will find that it is a double-walled sac with an opening into it. Insert the little finger of the left hand dorsal to the **ventral mesentery** or **lesser omentum** at the concave side of the stomach; push your finger caudalward, dorsal to the **pyloric stomach** and into the cavity (bursa) of the greater omentum. This opening is the **epiploic foramen.** The cavity within the greater omentum is sometimes called the **lesser peritoneal cavity,** or omental bursa, whereas the abdominal cavity caudal to or outside the

greater omentum is the **greater peritoneal cavity.** Between the stomach and spleen is the **gastrosplenic ligament,** which is the greater omentum.

When the stomach first forms in the embryo, it is in the medial position. Then it bends with its concave side ventrally. Following this the stomach turns, bringing its medial dorsal line to its extreme left, which is its greatest curvature. It is along this original medial dorsal line of the stomach that the greater omentum is attached. The shifting in the position of the stomach in embryological development accounts for the fact that more of the **right vagus nerve** supplies the dorsal surface of the stomach, whereas more of the **left vagus** supplies the ventral surface.

The **spleen** is a dark, elongate organ lying to the left of the stomach and supported in the greater omentum. It really lies between the two layers of the omentum. It has no ducts, and its products pass directly into the bloodstream. The spleen serves as a reserve blood supply in cases of severe bleeding. Cut off the greater omentum.

In the bend of the right side of the stomach, as shown in Figs. 3-2 and 3-3 is a thin, ventral mesentery known as the **lesser omentum.** It extends from the lesser curvature of the stomach to the liver. Anything entering the port of the liver is contained in the lesser omentum. The right free edge forms the epiploic foramen. Structures in the lesser omentum include the following. (1) The **common hepatic duct** (ductus choledochus) drains the liver; the **cystic duct,** which is short, soon joins it from the gallbladder, and the common hepatic duct continues to the duodenum. Before entering the diverticulum in the duodenum, the common hepatic duct is joined by the **pancreatic duct.** (2) The **main hepatic portal vein** is large and dark because of the coagulated blood or is blue because of injecting material. It has thin walls that are easily damaged. (3) The **hepatic artery** is probably injected red. It is small and usually lies between the two previous structures. Place your finger behind this area, tease away the connective tissue of the mesentery and omentum, and identify each of these. The **large postcaval vein** passes through the right dorsal part of the liver. The **hepatic veins** empty into the vena cava near the diaphragm. Blood from the hepatic artery and hepatic portal vein diffuses through the sinusoids of the liver into the tributaries of the hepatic veins, which join the postcava.

The **pancreas** lies along the duodenum and also extends along the caudal edge of the pyloric stomach in the greater omentum. It is often dark brown and varies in size, being smaller in older cats. The pancreas has two ducts. (1) The **main duct** may be found embedded in the pancreas parallel with the duodenum, entering it with the common hepatic duct. (2) The **accessory duct** is much smaller and opens into the duodenum two centimeters or more down the duodenum. The tributaries of the two ducts anastomose. If the blood vessels are not well injected within the pancreas, these ducts are difficult to distinguish. To dissect them turn the duodenum to the left side and dissect away the soft parts of the pancreas bit by bit over the fingertip, leaving the connective tissue within which the ducts are embedded. The thymus and pancreas are spoken of as the sweetbread.

Trace the **small intestine** from the pylorus of the stomach and determine its length. It is much longer than the body and thus has a tortuous course. It has no ventral mesentery caudal to the entrance of the common bile duct, but its **dorsal mesentery** is its principal support; through it pass the arteries, which bring nutriment and oxygen, and the veins, which absorb the nutriment from the digested food from the lumen of the intestine. The dorsal mesentery joins the dorsal body wall. The lymph capillaries are also in the dorsal mesentery close to the arteries and veins, but they are difficult to see. These absorb fat from the digested food and are called lacteals, since the contents are milklike in appearance. This fat is carried to the thoracic duct and eventually to the left external jugular vein or to the brachiocephalic vein.

Identify the divisions of the small intestines as follows: the **duodenum**, about three and one-half inches long and ∪-shaped, beside which is the pancreas; the **jejunum**, about eight inches long, which succeeds the duodenum; the remainder is the **ileum**, characterized by its many folds, which joins the large intestine. The transition from duodenum to jejunum is almost impossible to determine accurately from the external surface. When serial microscope slides are examined, they reveal glands in the wall of the duodenum and also foliate villi, whereas the jejunum has no glands of that type and in

the place of leaflike villi has filiform, or long and uniformly slender villi. The ileum has lymph nodules in its walls near the entrance to the large intestine. The ileum also has short, stubby villi. If microscope slides are available, these various structures may be pointed out by the laboratory instructor; otherwise their study may be omitted.

The principal glands that aid in digestion and their secretions are as follows: (a) the **salivary**; (b) the **gastric**, which produces gastric juice that contains hydrochloric acid; (c) the **pancreas**, which secretes strongly alkaline juice; and (d) the **liver**, which produces bile.

There are two kinds of thin, shiny **peritoneum**. (1) The **parietal** peritoneum is the thin, shiny inner layer of the thoracic and abdominal body walls. The kidneys, ureters, and gonads lie behind it; therefore, they are called **retroperitoneal**. (2) The **visceral** peritoneum is the thin outer layer of the stomach, liver, pancreas, intestines, and spleen and gives rise embryologically to the ventral and dorsal mesenteries, the broad ligament, and the greater omentum.

Spread out the small and large intestine and study the **dorsal mesentery.** It is really composed of two thin layers that are reflections of the **visceral peritoneum,** which forms the outer serous layer of the intestinal wall. The dorsal mesentery is the main support of the large and small intestines and functions quite well in the cat, whose body is in a horizontal position, but in man, whose body is in an erect position, it cannot support the intestine efficiently. The intestine in man settles down in the abdominal cavity, crowding the urinary and reproductive systems and complicating their function. This condition, with the accumulation of fat in the greater omentum, causes the abdomen to protrude.

Find the intestinal arteries, injected red, and the veins, blue or dark colored, lying between the two layers. Trace these from the intestine to the large mesenteric lymph node. This is a hard mass on the mesentery of the small intestines. Dissect it away and find the dark **superior mesenteric vein.** Refer to Fig. 4-4, where some of these structures of the digestive system are shown. Turn the intestine over to the right side of the cat and observe the **postcava** and the injected **dorsal aorta** lying close together near the middorsal line of the coelom. Find the left **suprarenal (adrenal) gland,** a small ovoid mass lying in the fat connective tissue close to the aorta and just cranial to the kidney (see Figs. 3-4 and 3-6). It is sometimes difficult to locate. The adrenals and kidneys are retroperitoneal organs, since they lie dorsal to, or behind, the peritoneum and do not lie in the coelom, or peritoneal cavity.

Find the beginning of the large intestine. The ileum enters it almost at right angles, and the part of the **ascending colon** that projects beyond the union with the ileum is the cecum. In herbivorous animals such as the rabbit, which is about the size of the cat, the cecum is as long and as large as your finger. In the cow it is as long and large as your arm. In carnivorous animals such as the cat, the cecum is quite short, and there is **no vermiform appendix.** In man the vermiform appendix varies in length from two to twenty-three centimeters, the average being from eight to nine centimeters. It is longest between ten and twenty years of age and tends to lose its size and lumen in older years. The evidence from various mammals indicates that the cecum degenerates to form the vermiform appendix.

Dissect away the small **lymph nodes** partially surrounding the **cecum** so that it may be seen clearly. The **ascending colon** is followed by the **transverse colon,** the **descending colon,** and the **rectum.** The large intestine of the cat is supported by a dorsal mesentery, whereas in man the transverse colon adheres tightly against the dorsal body wall. The beginning of the rectum corresponds with that of the first sacral vertebra, and on each side close to the anal opening is an **anal gland.** These are also known as scent glands in many mammals and are especially well developed in the skunk.

The **urinary bladder** is an oval-shaped organ caudal and ventral to parts of the intestines (see Figs. 3-1, 3-2, and 3-4). Its size varies, depending on the amount of urine it contains. Extending from it to the umbilicus is a **suspensory ligament** called the **ventral,** or **middle, ligament** of the bladder. In the fetus, it is the **urachus,** which is a continuation of the **allantois.** The **urethra** drains the urinary bladder to the exterior. The reproductive organs will be considered later.

THORACIC ORGANS (Figs. 3-1 to 3-3)

The dorsal and ventral mesenteries are represented in the thoracic cavity by two widely separated layers of splanchnic mesoderm, or visceral pleura, called the mediastinum. The **mediastinum** is a septum, or partition. It includes the space between the two pleural sacs, or **pleural cavities.** Into this space the **thymus, heart,** and lower ends of the **trachea** and **esophagus** migrate into position in the embryo. From the lower end of the trachea each lung bud pushes laterally and branches, forming a lung, which is covered by the visceral pleura of the mediastinum. The lateral wall of the mediastinum also forms the epicardium of the heart, while the pleuropericardial folds give rise to the **pericardium.**

The **thymus** is often darker than the other organs and lies cranial to the heart, within the mediastinum, between the cranial portions of the two lungs. It is elongate and irregular in shape and often bifurcates, since it arises from the two sides of the pharynx, or the epithelial lining of the third pharyngeal pouch. The size varies greatly in different cats, being smaller in the older ones. Remove the thymus intact. The mammary arteries and veins supply the thymus and pass along the dorsal side of the sternum. Branches pass through the body wall and supply the cranial mammary glands. The heart is surrounded by a sac, the **pericardium,** and it may be more or less covered with fat. Remove any fat and cut through the pericardium; pull the pericardium from over the heart. The coronary arteries are usually injected with red latex and are on the walls of the heart.

The **right lung** has four lobes (Figs. 3-2 and 3-3): **cranial** (apical), **middle** (cardiac), **caudal** (diaphragmatic), and **accessory** (intermediate). The **left lung** does not have the accessory lobe. The right lung is therefore larger than the left one, and the mediastinum is pushed somewhat to the left. The caudal vena cava lies to the right in the **caval fold,** a special fold of pleura between the caudal and accessory lobes. In general, each lung in mammals has a cardiac notch in the ventral border between the cranial and middle lobes. The left cardiac notch is usually larger since the heart tends to be more prominent on the left. In the cat there is a definite notch in the right lung, but the lobes of the left one do not extend as far ventrally and do not really form a notch but a cardiac space for the heart.

Observe the **cranial vena cava** entering the right atrium and the **caudal vena cava** coming through the diaphragm, and also entering the right atrium (Fig. 3-2). The **dorsal aorta,** or **main aortic arch,** curves toward the vertebral column. It is injected with a colored solution.

Examine the veins, arteries, and bronchi that pass through the hilus of the lung. The pulmonary arteries are usually injected blue. The bronchial tubes and bronchioli have the most definite walls and may be seen more clearly by mashing the lung slightly and washing away the softer tissue with water. Dry and examine with a hand lens.

There are two pairs of important nerves that pass down through the thoracic cavity.

1. The **phrenic nerve** courses ventral to the hilus of the lungs beneath the visceral peritoneum or pleura. The roots of the phrenic nerve arise from the fifth and sixth cervical spinal nerves in the cat but from the third, fourth, and fifth in man. The phrenic nerve is within the pleural folds in Fig. 3-1.

2. The **vagus** (the tenth cranial nerve) of each side emerges from the jugular foramen to course along the common carotid artery with the sympathetic trunk. It enters the thoracic inlet and passes dorsal to the hilus of the lung where it divides into dorsal and ventral esophageal branches. The branches of the right and left nerves join to form dorsal and ventral esophageal nerves to continue through the hiatus esophageus to the abdominal viscera. The recurrent nerve to the larynx arises from the right vagus at the axillary artery and from the left vagus around the aorta.

SOME DIFFERENCES IN INTERNAL ORGANS OF THE CAT AND MAN

1. Man has a vermiform appendix, whereas the cat has none.

2. There are six or more pyramids in the kidney of man, whereas in the cat there are fewer.

3. The large lymph node (mesenteric) located in the mesentery of the small intestines, in the cat, is not correspondingly large in man.

4. The central tendon of the diaphragm of the cat is almost circular, whereas in man it is greatly flattened dorsoventrally, forming right and left leaflets.

5. The liver of man has four lobes: the right, left, quadrate, and caudate, or spigelian. The liver of the cat has five lobes: the left medial, left lateral, right medial (cystic), right lateral, and the caudal (with a papillary portion).

6. The pancreas in man consists of a head, neck, and tail and lies along and behind the greater curvature of the stomach. In the cat it is relatively longer and more irregular in shape.

7. The lungs of man consist of a right lung with three lobes and a left lung with two lobes, whereas the cat has a right lung with four lobes and a left with three, not counting the incomplete divisions.

8. The ascending and descending colons in man no longer have mesenteries.

Name _____

Date _____

REVIEW QUESTIONS ON INTERNAL ORGANS WITH THE EXCEPTION OF UROGENITAL ORGANS

1. Of what kind of muscle does the diaphragm consist?

2. Locate the gallbladder and name the two ducts that help drain it.

3. What is the exact location of the thymus gland in the adult cat? (Fig. 3-2)

4. Of what does the greater omentum consist and what are its attachments?

5. What is the urachus?

6. How does it happen that more of the left vagus nerve supplies the ventral surface of the stomach and the right, more of the dorsal?

7. What organs constitute the viscera.

8. Locate the falciform ligament.

9. State what is meant by an organ being retroperitoneal; give three examples.

10. Name and locate the types, or classifications, of peritoneum.

11. Explain how the greater omentum is a modified part of the dorsal mesentery.

12. What is a diaphragmatic hernia and how does it arise?

13. Are the liver, stomach, spleen, and pancreas really within the peritoneal cavity? Why?

14. Explain the mediastinum of the thorax and name the structures it contains.

15. Explain the structure of the diaphragm and its central tendon.

16. Locate and explain the epiploic foramen.

17. What is the principal function of the spleen?

18. Does the cat have a vermiform appendix? Explain.

19. What is the mediastinum of the thorax, and what does it contain?

20. What organs are sometimes called sweetbreads?

GENERAL UROGENITAL SYSTEM

Fat is usually deposited near the kidneys. Remove this fat and also that surrounding the urinary bladder and covering the ureter. Dissect out the ureter on the left side. It is behind the peritoneum and therefore retroperitoneal. If the greater omentum contains excessive fat, it may be cut off close to its attachment to the stomach and spleen and discarded. Fat is usually deposited around the neck of the **urinary bladder.** Remove this fat by pulling it away with the fingers, being careful not to injure the **internal iliac (hypogastric) artery** and **vein** on each side, which extend to the bladder. If you have a male specimen, separate the spermatic cord at the place where it passes forward from each testis along the surface of the muscles lateral to the pubic symphysis and enters the abdominal cavity through the **inguinal canal.** Find the ridge of the **symphysis** by pressing with the finger; then cut to separate the pelvic bones, if this has not been done, being careful not to injure the sperm ducts. The urogenital systems of the male and female are constructed on the same general plan. Most of the fully developed structures have corresponding or homologous vestigial organs in the opposite sex. The organs in each sex consist of two principal groups.

1. The **urinary system** consists of the kidneys (metanephroi), ureters, urinary bladder, and the urethra; the pronephros and mesonephros of embryonic development have mostly disappeared.

2. The **reproductive,** or **genital, system** is composed of the ovaries or the testes with their accessory glands and ducts. The sexes may be distinguished by the external genitalia and by the location of the gonads in the adult. The ovaries are located just caudal to the kidneys, whereas the testes are in the scrotum outside the body. If your specimen is a male, the scrotum may have been removed with the skin, and the testes will be exposed in the anal region. Each student will be required to make a detailed study of the urogenital system of both sexes.

FEMALE UROGENITAL SYSTEM (Fig. 3-4)

If you happen to have a male specimen, perhaps later you can observe the following structures on some female cat in the laboratory that someone else is dissecting. If your specimen is a male, turn to the discussion of the male urogenital system, which immediately follows that of the female system.

1. The **female urinary system** consists of the following structures. Locate the **kidneys** and the **suprarenal (adrenal) glands** in the lumbar region. The two names for these glands have come about by the fact that in some animals they lie **against** the kidneys, so they are called adrenals. In man they are called suprarenals. However, the terms are often used interchangeably.

Pull the fat away from the kidneys, ureters, and bladder with the fingers. Observe that the kidneys are **dorsal** to the peritoneum, hence they are called **retroperitoneal.** The kidneys vary considerably in their relationship to one another, but the right is usually more cranial in the cat, while the reverse often occurs in man. A single artery goes to each kidney, while usually two veins drain the cat's right kidney. Carefully dissect the **parietal peritoneum** from the left kidney and its ureter. Occasionally man has two ureters from one kidney, which is an **anomaly.** This is not known to occur in the cat.

Cut the left kidney in half horizontally and observe its cavity, the **calyx,** or pelvis. The lining of the pelvis of the kidney is a continuation and enlargement of the wall of the ureter, and it in turn is a diverticulum, or outgrowth, of the **mesonephric (wolffian) duct** of the embryo. The outer portion of the kidney is the **cortex,** and the **medulla** is deep to it next to the calyx, consisting largely of radiating **uriniferous tubules** forming pyramids, a portion of which project into the pelvis as **papillae.** In the cat there is usually only one papilla in each kidney, whereas in man are several in each kidney.

Urine is separated from the blood by the process of **diffusion** from the **glomeruli.** These small, oval masses of blood capillaries, plus epithelial cells and capsular spaces (renal sinuses) constitute renal corpuscles. The nitrogenous waste of the urine is a by-product of **metabolic activity** of every living cell. It is absorbed and transmitted by the blood to the kidneys, where it diffuses from the blood into the capsular spaces and into the uriniferous tubules mentioned previously. From there it goes into the calyx, ureter, and bladder and is eliminated by the urethra.

The urinary bladder is covered with peritoneum cranially but is extraperitoneal or retroperitoneal caudally. Observe the place where the ureter enters the bladder and, in the male, the loop of the ductus deferens around it. The bladder varies in size, depending on the amount of urine it contains. The urethra of the female

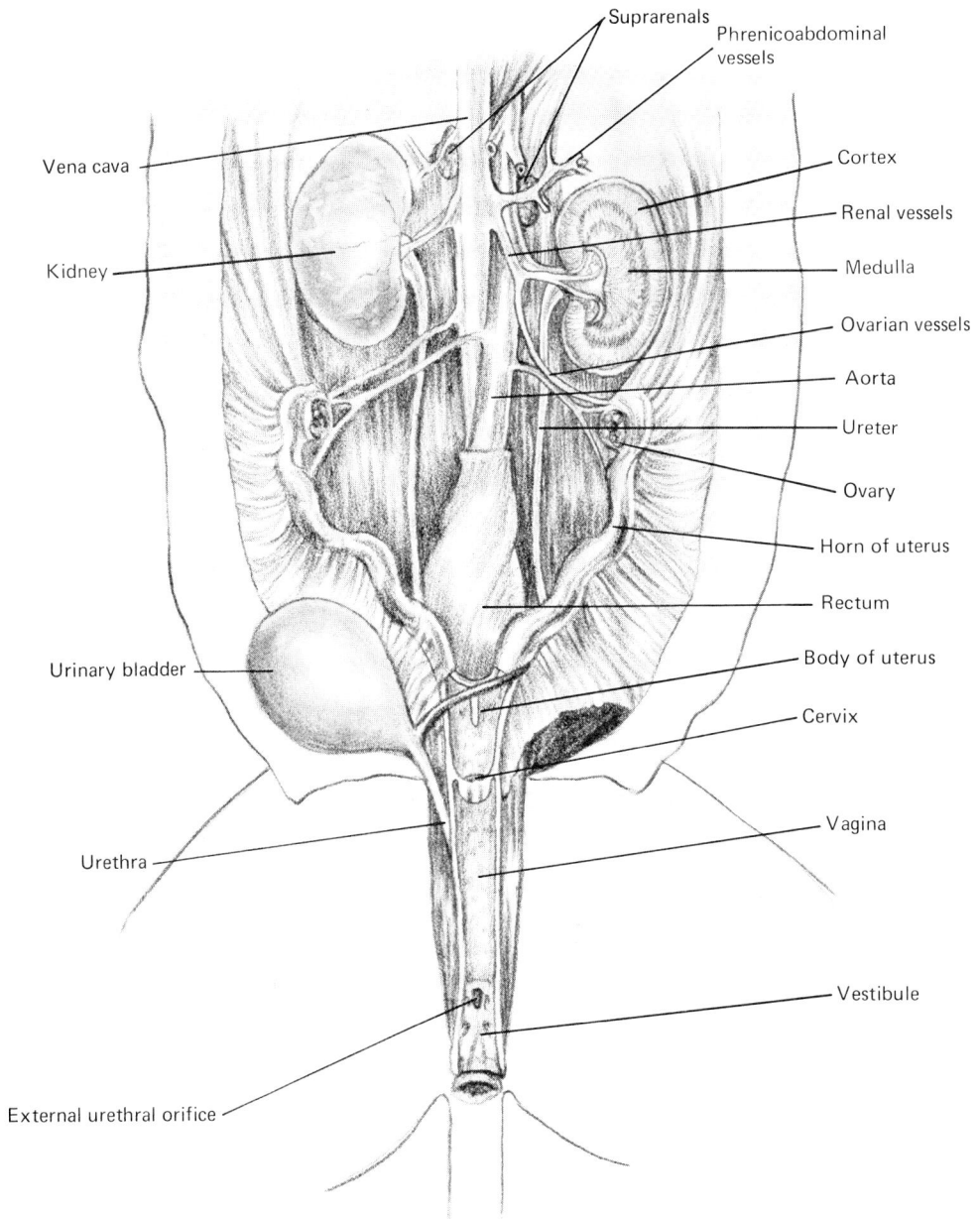

Fig. 3-4. Female urogenital system, ventral view.

drains the urine into the vestibule. Embedded in the urethral walls of the female are two sets of glands, the **paraurethral glands** and the **vestibular glands.** These are small and cannot be seen with satisfaction in gross dissection. The urethra arises embryologically from the ventral portion of the cloaca, whereas the dorsal part of the cloaca contributes to the extension of the rectum to the anal opening.

2. The **female reproductive system** consists of the following structures. The **ovaries** are slightly larger than a grain of rice and are suspended by the **mesovarium.** At the cranial end of each **cornu** or **horn, of the uterus,** next to the ovary, is the **fallopian (uterine) tube,** which is too small to be seen accurately. It is supported by the mesosalpinx. Eggs from the ovary enter the open end of the fallopian tube. The **horn of the uterus** extends caudally and meets its mate from the opposite side; the two merge together forming the **body of the uterus.** The uterus is supported by the mesometrium. Collectively, the

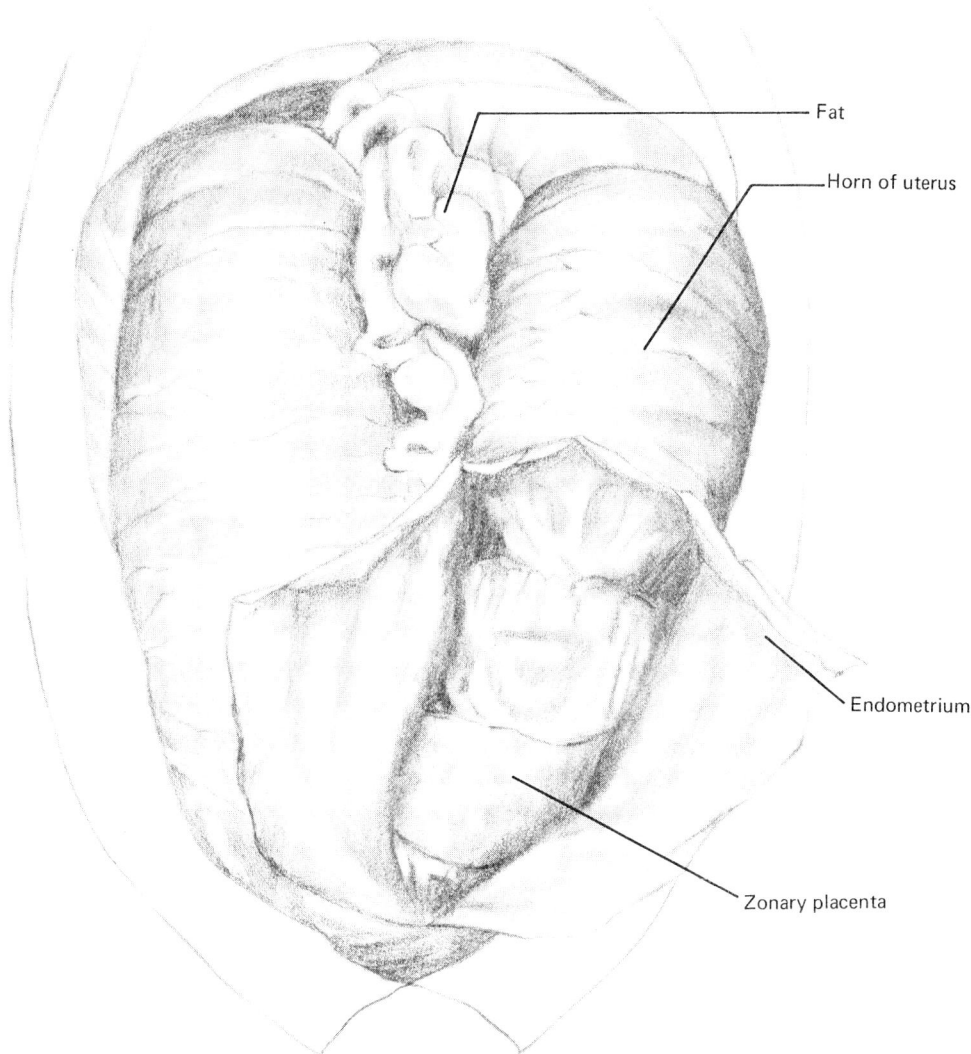

Fig. 3.5. Uterine horn with fetal membranes exposed.

supporting peritoneum for the genital apparatus is called the **broad ligament.** It consists of a double layer of parietal peritoneum from the body wall. Extending caudally and laterally from about the middle of each horn is the **round ligament.** Its caudal end is attached to the body wall in the approximate position of the inguinal canal. The **cervix** is the constricted area of the **body of the uterus** and projects into the **vagina.** Dissect these parts out and, if the cat is not pregnant, split them longitudinally. Observe the ovarian and uterine arteries in the broad ligaments. Each student in the laboratory should examine the urogenital system of the male also.

PREGNANT CAT (Fig. 3-5)

If your cat happens to be pregnant, you should be able to observe the principal organs involved. The gestation period for a cat is fifty-six to sixty-three days, and comparatively few female cats are obtained for dissection that are in the advanced stages of pregnancy.

1. **Loculi** are enlargements or swellings of the cornu, or horn, of the pregnant uterus, and by counting these, you can definitely determine the number of embryos, or developing young. The cat usually gives birth to four or six young in a litter. The size of each **loculus** is determined by the stage of development, or size of the embryo, and the placenta that it contains. All the

young of each litter are about the same size at any one time.

2. Remove one of the loculi by cutting through the uterine wall. The cut edges of the layers of the uterine wall can now be identified. (a) The **serosa** is the outer layer, which is thin and continuous with the **broad ligament** and the peritoneal lining of the body cavity. (b) The **muscular layer,** which is also thin, is next, but the fibers can be easily seen. (c) The **endometrium,** or **mucosal lining,** is mostly thin but quite thick in the area of the **zonary placenta,** which is the broad band surrounding the center of the embryo. When one side of the loculus is removed, part of the endometrium is divided, part comes off, and part is left with the embryo. Through these layers **uterine arteries** bring nutriment and oxygen to the placenta. When the uterine wall is pulled away from the placenta, some of these blood vessels are broken at the places where they enter the placenta against the chorion frondosum part of the chorion.

Observe the **maternal (uterine) blood vessels** that supply the various loculi. These enlarge during pregnancy, and their capillaries penetrate the walls of the uterus and come to lie in contact with the chorion frondosum to nourish the developing young.

The embryonic, allantoic, or umbilical blood vessels come from the embryo through the umbilical cord and terminate against this same chorion frondosum. Thus the blood vessels of the mother and the embryo come to lie very close to each other but are separated by the **chorion frondosum. Nutriment** and **oxygen** from the uterine arteries diffuse through this membrane into the embryonic, allantoic, or umbilical veins, while waste in the form of carbon dioxide and nitrogenous compounds diffuses from the embryonic, allantoic, or umbilical arteries into the uterine, or maternal, veins.

3. There are really four **embryonic membranes** (zonary placentas) in all vertebrates above the amphibians, and they will be considered here in order, beginning next to the inner surface of the wall of the horn of the uterus. (a) The **chorion** is the outermost layer of the embryonic membranes. If this layer is slit open near one end of an embryo and the cut edge is examined carefully, you may be able to separate what appears to be one layer into two. The chorion extends into and becomes

a part of the placenta. In the placenta it is called the **chorion frondosum** and is the most significant part of the placenta, because it is through this membrane that all nutriment and waste products must diffuse in reaching or leaving the embryo. The chorion frondosum is thrown, or develops, into projections known as villi. (b) The **allantois** is divided into an outer part, which is here fused with the chorion and which you may have just separated, and an inner layer, which is still closer to the embryo. The wall of the allantois contains the embryonic or umbilical blood vessels, whose capillaries come close to the uterine, or maternal, blood vessels in the villi of the placenta. (c) The **amnion** is the third embryonic membrane and is a very thin layer next to the skin or fur of the fetus. It is very closely fused here with the inner layer of the allantois, so they probably appear as one thin, transparent layer. (d) The **yolk sac,** the fourth embryonic membrane, develops around the yolk of the microscopic egg and in the mammal dilates as the embryonic membranes develop. The wall of the yolk sac shrinks as its contained nutriment is absorbed, and it is left adjacent to the placenta, whereas its stalk is drawn within the body of the embryo during the first few weeks of gestation. The yolk sac will not be seen outside the body of the developing kitten unless the kitten is less than about one inch long. Parts of the chorion and allantois, with allantoic blood vessels, form most of the embryonic part of the placenta.

4. The fully developed **placenta** is a thickened band, or zone, surrounding the embryo, and because of its shape in the cat is called a "zonary placenta." The placenta is composed of two principal parts. (a) The **maternal part** consists of a portion of the mucosal lining (endometrium) of a loculus. This area is thickened to a rather broad band at places where the villi are located. The uterine, or maternal, blood vessels penetrate the serosa and muscular layers of the loculus of the uterus to reach the placenta proper. (b) The **embryonic part** consists of a portion of the chorion known as the chorion frondosum and a part of the allantois where its umbilical or allantoic blood vessels end against the outer limits of the villi. The walls of the villi constitute the chorion frondosum. The amnion does not enter into the placenta but forms a thin, transparent sac, or covering, for the

embryo and becomes continuous with the outer limits of the umbilical cord.

Remember that no blood vessels or blood passes from the mother to the embryo or from embryo to mother. Only oxygen, nutritive substances, and a few disease-producing organisms are known to pass to the human embryo, whereas only the carbon dioxide and nitrogenous wastes pass from the embryo to the mother's blood. The embryo has to produce all of its own blood. The embryonic, umbilical, or allantoic blood vessels may be seen on the under or embryonic side of the placenta.

5. The **umbilical cord** is the connection extending from the inner surface of the placenta to the abdominal wall of the embryo, and the umbilical or allantoic arteries and veins may be seen extending through it.

Immediately after the birth of each kitten the placenta loosens, sloughs off from the uterus, and is expelled. The mother cat gnaws the umbilical cord in two, thus freeing the newborn kitten from the afterbirth. Soon after the birth the proximal end of the umbilical cord dries up and sloughs off beyond the umbilicus, which is the scar on the outer surface of the abdominal wall of the kitten. Soon after the birth of each of the kittens the afterbirth is expelled.

SOME COMPARISONS REGARDING GESTATION AND BIRTH

The female cat gives birth to from two to eight young in a litter, while the human normally produces one. One embryo is produced for each egg in the cat. In the female human, when more eggs are available, there may be more embryos; moreover, the fertilized egg may divide, forming, multiple embryos such as identical twins or triplets. In the cat the young embryos develop out in the **cornu,** or **horn,** of the uterus, while in man this occurs in the **body** of the uterus, since there are no horns. In all cases of multiple birth, the size of each young at birth depends on the amount of blood going to the **loculus** where each develops. The ends of the **cornua** have the least amount of blood. In the case of pigs, in whom at least eight piglets are usually produced in a litter, there is often much variation in size of the young at birth. The gestation period for cats is **fifty-six** to **sixty-three days;** in the pig it is **seventeen weeks,** and in man, it is usually about **thirty-eight weeks.** The cat has a **zonary placenta,**

which surrounds the developing young. The **villi,** which are a part of the **chorion,** are, at first, all over the chorion but later disappear, with only a band, or zone, remaining. During early embryonic life man has the villi all over the chorion, but later the villi disappear, except for a circular area, which at full term is the size of a six-inch pancake and as thick as the palm of your hand. Here we have variation in the distribution of the villi on the placenta of man, which is interpreted as "recapitulation." Recapitulation supports the theory of evolution.

MALE UROGENITAL SYSTEM (Fig. 3-6)

The **urogenital system** consists of the **urinary** and the **reproductive** systems. The kidneys, ureters, urinary bladder, and urethra form the urinary system. Cut away the **peritoneum** covering the left kidney and open it horizontally, as in the female. The peritoneum is usually only ventral, since the kidneys are retroperitoneal. The kidney itself has a thin connective tissue capsule. The cavity near the center is the **calyx,** or **pelvis,** and the outer portion consists of the **cortex,** which is near the outer surface, and the **medulla,** which is next to the calyx. The **ureter** and the kidney are retroperitoneal. Trace the ureter to the **bladder,** where it is surrounded by a **loop of the ductus deferens** before it enters the bladder. The bladder is supported by a ventral and two lateral peritoneal ligaments. The ventral one contains the **urachus** in the fetus and the lateral ones the umbilical arteries.

Separate the **symphysis,** or union of the two **pelvic bones,** if this has not been done, and dissect out the urethra from the urinary bladder to the base of the **penis.** The urinary bladder has a long narrow neck that joins the **urethra.** At its beginning the urethra is surrounded by the **prostate.** Here the ductus deferens from each side enters. On each side of the urethra near the penis is a small **bulbourethral gland.** Cut away the skin and scrotum that surround the testes. The testes are retroperitoneal during their formation and migration into the scrotum. Near the testis is a mass of tubules, the **epididymis,** which continues as the **ductus deferens** (**sperm duct**), which in turn passes forward under the skin and enters the body cavity through the **inguinal canal** opening. The wall of the **scrotum** is really an outpouching, or diverticulum, of the abdominal wall and is lined by an extension of peritoneum, **tunica vaginalis.**

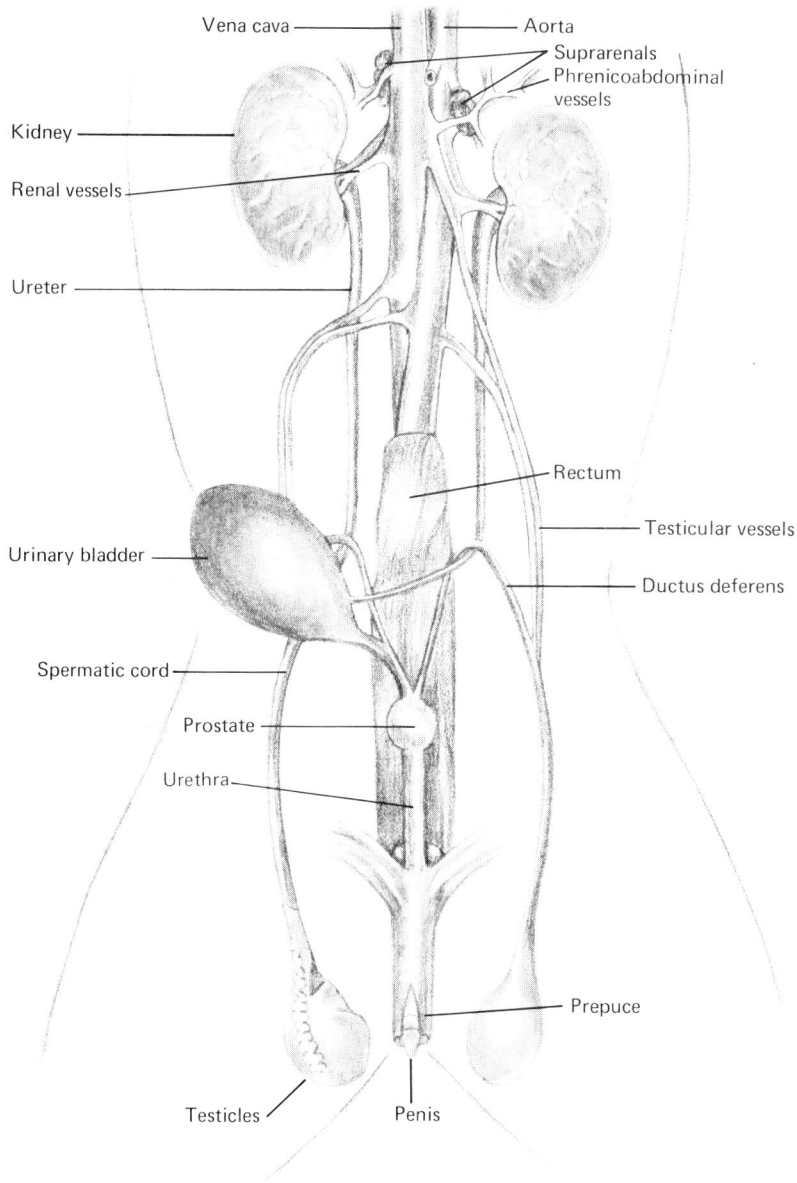

Fig. 3-6. Male urogenital system, ventral view.

Sometimes the abdominal wall weakens at the inguinal canal in the human male, and the intestines protrude down into the scrotum, producing an **inguinal hernia.** A quite similar condition occurs in the human female, but this is associated with the femoral artery. An operation usually corrects either of these defects.

Trace the ductus deferens within the abdominal cavity and find the **loop** around the ureter. This loop occurred because the testis in the embryo was close to the kidney; it migrated down ventral to the ureter and took its blood vessels with it. Near the kidneys the **internal spermatic (testicular) artery** and **vein** branch

from the dorsal aorta and caudal vena cava and pass down through the **inguinal canal.** As the vessels travel to the **inguinal ring,** they join with the ductus deferens as part of the **spermatic cord,** which thus includes the **testicular vessels, nerve, lymphatics, ductus deferens,** and the **peritoneal extension.**

DEVELOPMENT OF UROGENITAL SYSTEM

The embryological development of the **kidney** of cat or man is quite complicated, since there are what might be called three "kidneys," or **three main parts** for each kidney. These three parts are the (1) **pronephros,** or head kidney;

100

(2) **mesonephros**, or middle kidney; and (3) **metanephros**, or kidney of the adult animal. The first two disappear, for the most part, since in the adult only vestigial remnants, which are not easily seen in gross dissection, remain. See the accompanying chart, p. 102.

The **ductus deferens** has an unusual history. When it is first formed in the embryo, it is called the **wolffian duct**, or **mesonephric duct**, and drains the middle kidney, or **mesonephros**. When the permanent kidney, or **metanephros**, forms in the older embryo, the mesonephros almost completely disappears.

Some of its uriniferous tubules remain and connect with the **testis** and become the **efferent ductules of the epididymis** in the male or the **epoophoron** in the female. Other uriniferous tubules become vestigial. The wolffian duct becomes the **ductus deferens** for the passage of sperm. In the female the **wolffian duct** becomes the vestigial **Gartner's duct**. Since the urethra is a duct for both urine and sperm in the male, it functions as a urogenital duct, but it is only urinary in the female. The **epididymis** consists of the convoluted portion of the efferent ductules, or mesonephric tubules, and of the ductus deferens by the side of and against the testis.

HOMOLOGIES IN UROGENITAL SYSTEMS OF THE CAT AND MAN

The urogenital organs of the male and female in the cat and man are built on the same general plan. In the young embryo the sex organs are undifferentiated and hence are given **generalized terms**, as in the middle column of the accompanying chart. Organs having the same origin and general structure are said to be **homologous**. Here we often see the same generalized structure of the early embryo becoming different organs in the adult male and female.

Sometimes the male or female reproductive organs do not differentiate completely into a typical anatomical male or female but become partly developed reproductive organs of both male and female in one individual. This occurs in many groups of animals, including the human being. Such individuals are called **hermaphrodites** or **intergrades**. Man's body does not always follow a certain strict standard anatomical pattern.

It is interesting to know that in the embryological development of the cat or man there is a **brief summary** or **review** of the structure or kinds of the kidneys of various groups of lower animals. In the *myxinoid Cyclostomis* and a **few bony fishes**, the **pronephros** is the functional kidney in the adult, and the mesonephros and metanephros do not form. The **lamprey, most fishes**, and the **amphibians** have the **pronephros** in their early development but only the functional **mesonephros** as adults. **Reptiles, birds**, and **mammals** have the **pronephros** in early embryonic development, the **mesonephros** in later embryonic development, but only the functional **metanephros** as adults.

Thus, when the kidneys pass through stages in their development that are similar to those of the adults of lower vertebrates, as the preceding facts show, it is known as **recapitulation**. These facts are interpreted by most embryologists and anatomists as supporting the theory of **evolution**.

SOME DIFFERENCES IN UROGENITAL SYSTEMS OF THE CAT AND MAN

1. The female cat has a bipartite uterus, whereas the female human has a simplex uterus.

2. The right kidney of the cat is located slightly farther forward than the left kidney, whereas in man the left kidney is slightly higher in the body than the right.

3. Usually two renal veins drain the right kidney of the cat. This condition is found less often in man.

4. Usually six or more medullary papillae, or pyramids, are found in each of man's kidneys, whereas the cat usually has only one.

5. The uterine tubes are relatively much longer in the female human than they are in the cat.

6. The ovaries of the cat are located relatively closer to the kidneys than they are in man.

7. The ductus deferens of man enters the urethra immediately below the bladder, where the prostate gland is located. In the cat this area seems farther caudal, since the neck of the bladder is hardly greater in diameter than the urethra.

8. The seminal vesicles are well developed in man but are poorly developed or absent in the cat.

9. The placenta of man is discoid, whereas in the cat it is zonary.

10. In the male cat an inguinal hernia seldom, if ever, occurs, whereas in man it is not unusual. Its common occurrence in man is partly because of the erect position of the body.

Comparable urogenital organs

Adult male	Young embryo	Adult female
Disappears	Pronephros	Disappears
Efferent ductules	Mesonephros	Epoophoron
Kidney	Metanephros	Kidney
Ureter	Diverticulum of wolffian duct	Ureter
Urinary bladder	Parts of cloaca and allantoic stalk	Urinary bladder
Urachus	Part of allantoic stalk	Urachus
Urethra	Ventral part of cloaca	Urethra
Testes	Gonads	Ovaries
Sperm duct, or ductus deferens	Wolffian duct	Gartner's duct (is degenerated)
Vagina masculinus, or prostatic utricle (is degenerated)	Müllerian duct	Fallopian tube, uterus, and vagina
Prostate gland	Urethral glands	Paraurethral gland
Bulbourethral gland	Urethral glands	Vestibular glands
Penis	Phallus	Clitoris
Scrotum	Muscles and skin	Labium majora

REVIEW QUESTIONS ON UROGENITAL SYSTEMS

1. What parts of the urogenital system are retroperitoneal?

2. What duct conducts germ cells from the testis? the ovary?

3. How long is the gestation period of the cat? of man?

4. What is the horn of the uterus and how does it adjust itself to the developing embryos?

5. Name the four principal embryonic membranes of the pregnant cat.

6. Name the embryonic and maternal membranes that are most involved in the formation of the placenta.

7. What happens to the embryonic membranes when the young are born?

8. Describe how nutriment within the maternal, or uterine, artery finds its way into the bloodstream of the embryo.

9. How does the nitrogenous waste within the circulation of the embryo reach the blood vessels of the mother cat?

10. Explain the term mediastinum. (See definitions of terms.)

11. Name the structures that are homologous in the adult male and female urogenital systems.

12. Why is it that the urethra is strictly a urogenital duct in the male but is not in the female?

13. Give in the corresponding order the name of the duct, or tube, that drains the following structures: (1) ovary, (2) urinary bladder, (3) testis, (4) kidney, and (5) gallbladder.

14. Trace the passage of sperm cells, or spermatozoa, from the testis until eliminated from the body.

15. Name the duct that takes urine to the bladder and the duct that takes urine away from the bladder.

16. What is an inguinal hernia?

17. What are the main contents of the spermatic cord as it passes through and beyond the inguinal canal?

18. Are the testes of the cat or man within the body cavity or are they retroperitoneal?

19. What is the broad ligament and where is it located?

20. What are the three kinds or parts of the kidneys embryologically?

21. Why does the sperm duct loop around the ureter in the male?

22. Do embryos of the cat develop in the body or horn of the uterus? Where do embryos develop in in the human being?

23. What is a hermaphrodite?

24. Why do the testicular arteries and veins, which supply the testes, arise near the kidneys rather than in the pelvic region?

25. What constitutes the afterbirth?

FOUR
Venous and lymphatic systems of the cat

The **blood system** is a two-way system in that the arteries carry oxygenated blood laden with nutriment from the heart out over the body, where oxygen and nutriment diffuse through the walls of capillaries to supply the various cells, which in turn give off carbon dioxide and broken-down substances of metabolism. This waste material diffuses into the venous capillaries, which unite to form veins that return this blood to the heart. The **cranial** and **caudal venae cavae** are the two principal vessels entering the heart. The larger arteries and veins tend to run parallel with one another, but many variations exist as a result of their early formation in the embryo as a fine meshwork, or network, of capillaries. The blood begins to flow along a certain course, which is not always the same in a given area in different animals. These vessels that carry most of the blood enlarge, while others remain relatively small. Therefore blood vessels, particularly veins, in cats differ not so much in the area supplied or drained as in the way they connect with larger vessels. In **dissecting** or **surgery** one needs to know not only the places where blood vessels are usually located but also, if not in the expected location, the places where they are then most likely to be found in relation to adjacent muscles and nerves. Veins usually vary more in their positions and branchings than do arteries. As you dissect, compare your specimen with others in the laboratory and note these anomalies.

The **lymphatic system** is a one-way system in that lymph flows only toward the heart, or, to be more exact, toward the large veins that take the lymph with the blood to the heart. The **thoracic duct,** or main trunk, of the lymphatic system is shown in Fig. 4-5.

TRIBUTARIES OF CRANIAL VENA CAVA, OR PRECAVA (Fig. 4-1)

There is considerable variation in the veins of the cat. There is variation in the way they unite with one another and also in the way they are injected. Therefore the veins on your cat may not be exactly as in the following description. Cut off the thymus gland, previously identified, the fat anterior to the heart, and the pericardium, which is the membranous sac surrounding the heart. The **cranial vena cava** is the largest, dark-colored vessel entering the heart on its cephalic surface. It is formed by the union of the two brachiocephalic veins, and its principal tributaries are as follows.

1. The **azygos vein** is the tributary of the precava close to the heart. Pull the heart over to the cat's left side and observe the azygos vein on the right side close against the spinal column. It is large and arises close to the diaphragm. The tributaries of the azygos are the **intercostals** from the body wall. The azygos vein is only on the right side; the word "azygos" means unpaired.

2. The **internal thoracic vein** is the second tributary into the cranial vena cava. It drains the cranial mammary glands or comparable area in the male. It passes forward along the dorsal surface of the sternum and then into the mediastinum close to the thymus gland. It usually joins the one of the other side before emptying into the cranial vena cava.

TRIBUTARIES OF BRACHIOCEPHALIC VEIN (Fig. 4-1)

1. The right **costocervical vein** often enters the **brachiocephalic vein;** however, it may have a common base with the **vertebral** in the region

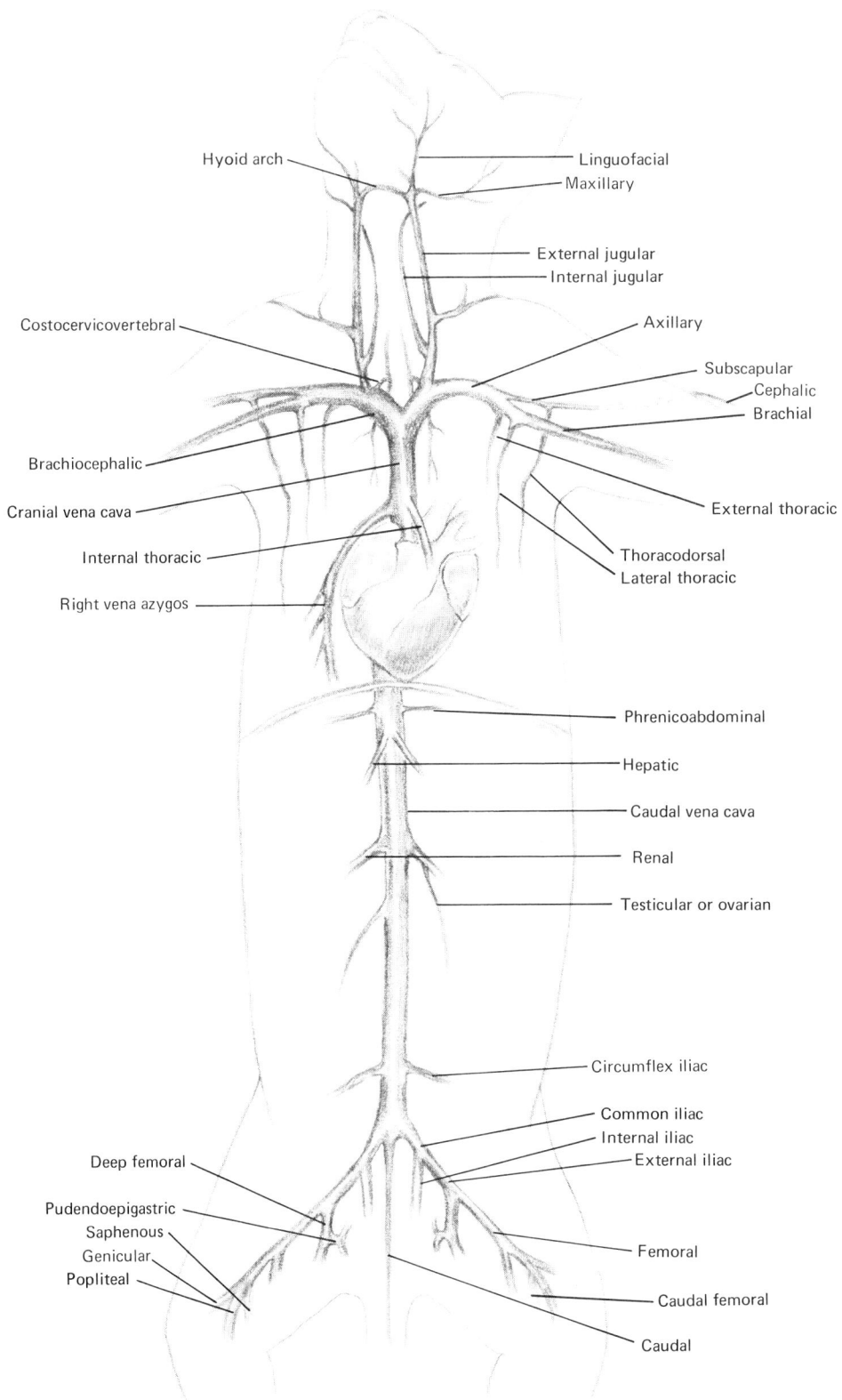

Fig. 4-1. Venous system.

of the first rib. Therefore it is mentioned in both groups. In this case they form the **costo-cervicovertebral** axis, which enters the **cranial vena cava** close to the **internal thoracic vein.** Dissect the right vertebral vein to the place where it emerges from the transverse foramen of the seventh cervical vertebra.

2. The **internal jugular vein** arises at the base of the skull and extends down the neck beside the trachea close to the **common carotid artery** and the **vagus nerve.** It usually enters the brachiocephalic vein but may enter more cranially into the external jugular vein.

3. The **axillary vein** comes from the brachial region of the front limb and receives the **subscapular vein,** the base of which may be almost parallel to the external jugular vein. The axillary joins the external jugular vein at the first rib to form the brachiocephalic vein.

4. The **external jugular vein** from the head and neck lies just under the platysma muscle on the ventrolateral surface of the neck. It joins the axillary on each side to form the brachiocephalic vein.

TRIBUTARIES OF AXILLARY VEIN (Fig. 4-1)

The term **subclavian** is listed in the nomenclature to include **median, brachial,** and **axillary** veins. The median vein of the forearm, or antebrachium, continues in the upper arm, or brachium, as the brachial vein, which then becomes the axillary vein. The axillary vein is more specifically the proxima part of the main vein from the forelimb under the clavicle. It unites with the external jugular vein to form the brachiocephalic.

Dissect out these veins and the tributaries of the axillary on the left limb as follows, beginning close to the entrance of the left external jugular vein. The tributaries of the axillary vein vary in different specimens, and the descriptions here will not be accurate for all cats.

1. The **subscapular vein** runs along the caudal border of the scapula and joins the axillary vein close to its origin.

2. The **external thoracic vein** is small and joins the axillary vein close to the first rib. It extends ventrally and drains the pectoral muscles close to the sternum. This vein and the two following vary considerably in the place where each joins the axillary vein. Seldom can each of the three be traced its full extent in the same cat.

3. The **lateral thoracic vein** arises along the inner surface of the pectoralis minor muscle near the outer edge of the mammary gland, runs forward along the medial side of the latissimus dorsi, and joins the axillary or sometimes the subscapular. After the lateral thoracic vein leaves the mammary region it is called the **external mammary.**

4. The **thoracodorsal vein** arises from the upper chest wall under the scapula and joins the axillary usually close to the **caudal circumflex humeral vein,** which is on the level with the lateral surface of the body wall.

5. The **cephalic vein** is on the cranial surface of the forearm and joins the axillary vein by passing behind the shoulder joint to join the subscapular. A **median cubital vein** connects the cephalic and brachial veins near the elbow.

Dissect each of these veins so that all may be demonstrated. Remember that the instructor can usually tell at a glance whether or not a serious attempt has been made to dissect them.

TRIBUTARIES OF EXTERNAL JUGULAR VEIN (Fig. 4-1)

Dissect away the fascia and muscles covering the right external jugular vein and identify the following veins.

1. The **transverse scapular (superficial cervical) vein** flows into the lower portion of the external jugular. It comes from the shoulder cranial to the scapula, where it receives blood from the **cephalic,** previously mentioned.

2. The **thoracic lymph duct** of the **lymphatic system** (see Fig. 4-5) enters the base of the **left external jugular vein** after passing up the thoracic region over the left side of the trachea and esophagus.

3. The **right lymphatic duct** (see Fig. 4-5) is short and inconspicuous. Look for it close to the base of the **right external jugular,** or it may enter the **brachiocephalic** or **axillary veins.** It is usually difficult to identify, but sometimes it is beaded and relatively large. It is believed to be a **remnant** of a right thoracic duct, a mate to the left.

4. The **hyoid arch vein** extends across from the opposite side near the hyoid cartilage and may have one or more median vein tributaries.

5. The **linguofacial vein** arises between the eye and the mouth by the union of the **frontal**

and **nasal veins**. It receives the **mandibular labial vein** near the angle of the mouth. The facial vein passes caudally close to the lymph glands and receives the **lingual vein**. It usually joins with the hyoid arch vein before merging with the maxillary vein.

6. The **maxillary vein** is formed by tributaries from the side of the head and region of the ear. The **caudal auricular vein** usually joins the **superficial temporal vein** below the parotid salivary gland to form the maxillary vein, which passes lateral to the submaxillary gland to join the **linguofacial vein**. The superficial temporal is formed by tributaries at the back of the eye and usually receives the **rostral auricular vein** and a deep subfacial or **internal maxillary vein**. Sometimes the two auricular veins have a common base before joining the superficial temporal.

TRIBUTARIES OF CAUDAL VENA CAVA, OR POSTCAVA (Fig. 4-1)

1. Dissect out the caudal vena cava from the right atrium to the diaphragm and find the **phrenicoabdominal veins** embedded in the cranial or caudal walls of the latter.

2. The **hepatic veins** are embedded deep in the liver and will be considered later in connection with the hepatic portal system.

3. The **renal veins** return blood from the kidneys. This blood has had most of its nitrogenous waste removed as it passed through the glomeruli of the kidneys. This waste will find its way as urine into the ureters and the urinary bladder to be eliminated through the urethra. Two renal veins are sometimes found on the right side and, more rarely, two on the left.

4. The **right testicular vein** of the male or the **right ovarian vein** of the female enters directly into the caudal vena cava, but on the left side more often each enters the left renal vein. Examine the specimen closely and note these differences. The testicular veins are often not well injected, since they are small and the injecting material must pass the valves in the wrong way.

5. The **circumflex iliac veins** drain the small of the back and are closely associated with the arteries of the same name.

6. The **common iliacs** are the two tributaries that unite in the pelvic region to form the caudal vena cava. The **right common iliac** receives the **caudal**, or **sacral**, **vein** from the tail. Sometimes there is also one on the left. Each

common iliac is formed by (a) the **external iliac (femoral) vein**, which is the large principal vein from the leg, and (b) the **internal iliac (hypogastric) vein**, which unites with the external iliac near the body wall. Sometimes the two common iliacs extend forward separately to the region of the kidneys, where they unite to form the caudal vena cava. When this occurs, it is considered an anomaly, or a marked deviation from the normal structure.

TRIBUTARIES OF EXTERNAL AND INTERNAL ILIAC VEINS (Fig. 4-1)

The **right** and **left common iliac veins** extend from the lower end of the caudal vena cava to the abdominal wall, where each is formed by the union of the long **external** and the short **internal iliac veins**.

1. Tributaries of the **external iliac vein** are as follows. The pelvic limb of the cat is to be dissected from the ventral or medial surface, but there are so many variations that the following descriptions may not be accurate in all details for any one cat. Veins are named according to the structures they drain—not in the way they unite with one another. In the thigh the external iliac continues as the femoral vein.

a. The **circumflex iliac vein** branches from the external iliac near the body wall and extends laterally.

b. The **caudal epigastric vein** comes from the groin on the inside of the ventral body wall and, in the female, from the caudal mammary glands.

c. The **deep femoral vein** has its base immediately beyond the caudal epigastric. It comes over the cranial surface of the adductor muscle and accompanies the deep femoral artery. Sometimes this vein is not injected.

d. Two or more **muscular branches** arise at about the center of the dorsal surface of the gracilis muscle and join the external iliac, which is now called the femoral, at about the center of the thigh. The femoral vein is parallel and usually medial or caudal to the femoral artery.

e. The **genicular vein** comes from the region of the kneecap and is close to the surface. It extends up the thigh to join the femoral vein.

f. The **medial saphenous (saphena magna)** is superficial, or next to the skin, coming up along the inner side of the calf of the leg and extending across the lower part of the gracilis muscle, accompanied by the saphenous artery and nerve to its union with the popliteal and genicular vein to form the femoral (external iliac) vein.

g. The **lateral saphenous vein** comes up the superficial caudal surface of the calf of the leg and joins the **caudal femoral vein,** which brings blood down from the caudal region of the thigh to form the **popliteal vein.** Other tributaries of the popliteal are the **cranial** and **caudal tibial veins.** Their bases may be seen as they join the popliteal and then may be traced down into the calf of the leg. (Quite often these veins are poorly injected.) The popliteal vein continues as the femoral vein, and the femoral vein becomes external iliac. Then the more distal branches of the internal iliac join the latter two veins.

2. Tributaries of the **internal iliac vein** are few in number. The internal iliac as such is very short. Expose its base near the body wall by reflecting the gracilis, semimembranosus, and adductor muscles and observe that its three tributaries unite close together. It ends as the **caudal gluteal vein.**

HEPATIC PORTAL SYSTEM (Fig. 4-2)

The tributaries of the **hepatic portal vein** drain the walls of the digestive tract and its mesenteries caudal to the **diaphragm.** It also drains the pancreas and spleen. The hepatic portal system consists entirely of veins that begin as capillaries associated with the aforenamed organs and ends in sinusoids in the liver. There is much variation in the veins that drain the **stomach, pancreas,** and **spleen,** and those shown in Fig. 4-2 may not be completely accurate for your particular specimen. Remember that veins are named from the organs they drain and not by the manner in which they unite with one another. This system is often not injected, and thus its smaller vessels are more difficult to identify. Following are the more important veins that unite to form the hepatic portal vein.

1. The **cranial mesenteric (jejunal) vein**

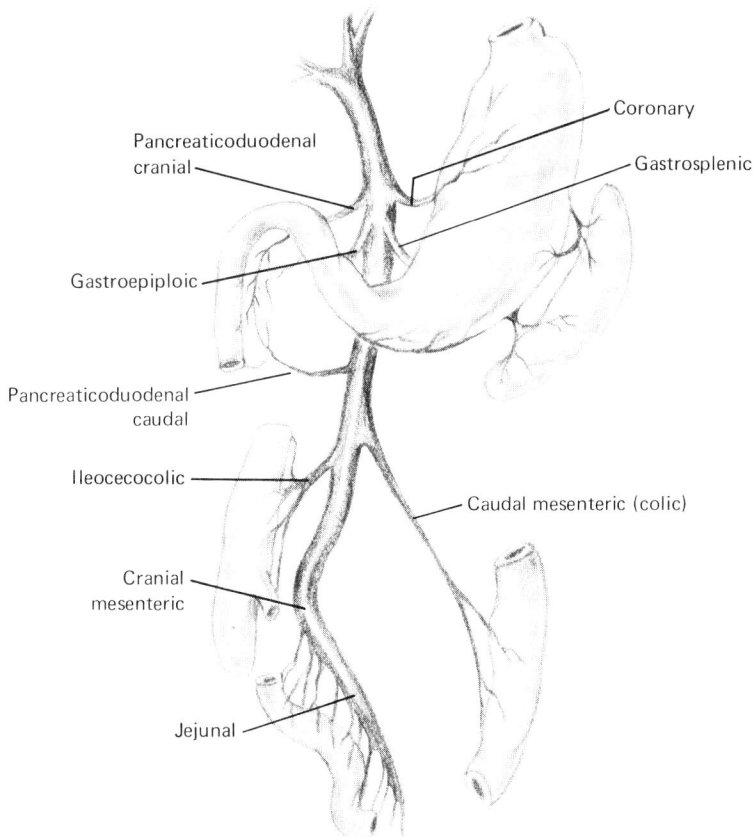

Fig. 4-2. Hepatic portal system.

arises in many small tributaries in the wall of the **small intestine** and **ascending colon.** Most of these vessels extend through the thin dorsal mesentery and unite with one another near the large mesenteric lymph node. Dissect it away and find the superior mesenteric vein.

2. The **caudal mesenteric (colic) vein** arises in the wall of the **descending colon,** extends forward through its mesentery, and joins the cranial mesenteric vein.

3. The **pancreaticoduodenal vein** is small and drains part of the **duodenum** and a portion of the **pancreas** adjacent to it. It can usually be seen on the proximal bend of the duodenum as it extends forward to enter the portal vein. The pancreaticoduodenal is often difficult to identify because it is wholly or partially surrounded by adjacent tissue. Examine the dorsal surface of the lesser omentum, where it is sometimes exposed with little or no dissection as dark or blue in color.

4. The **gastroepiploic vein** drains the posterior wall of the pyloric portion of the stomach, pancreas, and usually portions of the spleen and greater omentum. It passes forward dorsal to the pyloric opening of the stomach to join the portal vein, or sometimes it joins the pancreaticoduodenal vein. It is small and often difficult to trace.

5. The **coronary vein** is small and drains the region of the lesser curvature of the stomach close to the cardiac opening. It extends through

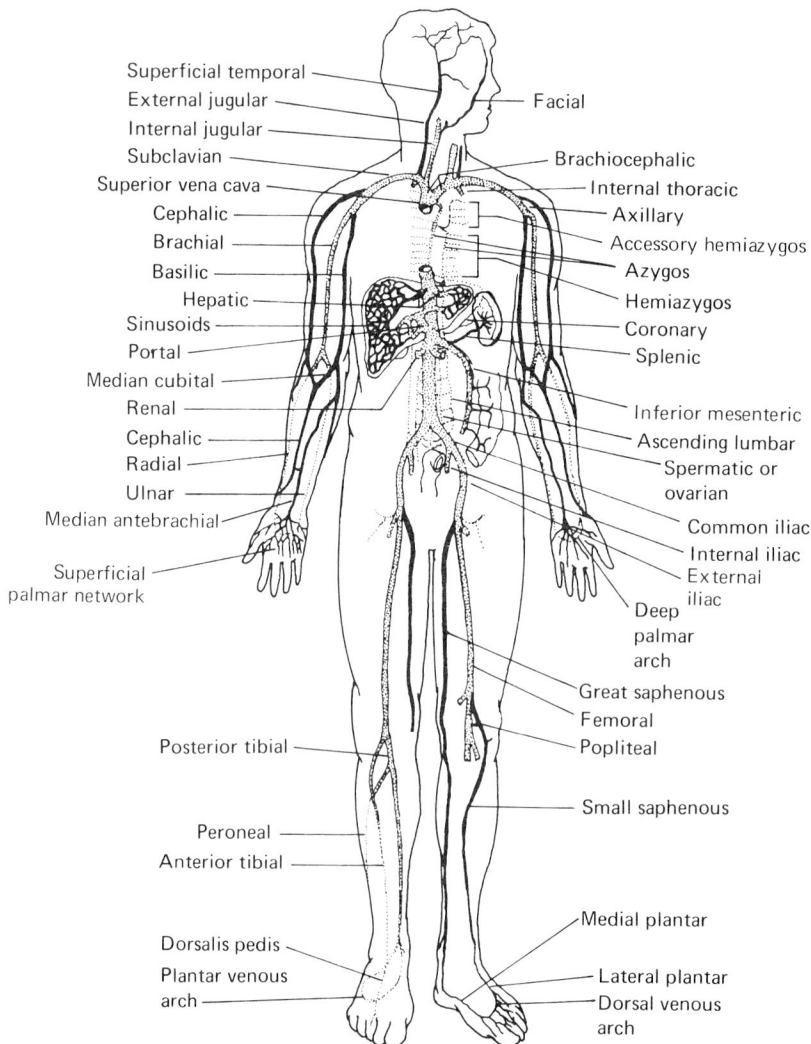

Fig. 4-3. Human venous system. (From Millard, N. D., King, B. G., and Showers, M. J.: Human anatomy and physiology, Philadelphia, 1956, W. B. Saunders Co.)

the lesser omentum to join the gastrosplenic vein, or sometimes it connects directly with the portal vein.

6. The tributaries of the **gastrosplenic vein** drain most of the spleen and dorsal wall of the stomach as they pass across its dorsal side to unite into one vein before joining with the common mesenteric to form the large common hepatic portal vein.

The **tributaries of the hepatic portal vein** absorb carbohydrates and proteins, largely from the walls of the small intestines. These nutrients are transported to the liver by the large **hepatic portal vein,** where it branches and rebranches, ending in intralobular veins and portal sinusoids between the radiating cells of the liver lobules. From here the blood enters the **intralobular** (central lobular) **veins,** which are tributaries of the **hepatic veins,** and hence into the **caudal vena cava.** The sinusoids are in contact with the liver cells, which absorb the excess carbohydrates in the blood and store them in the form of glycogen. Later this glycogen is "doled out" as needed into the intra-

lobular veins, which connect with hepatic veins, the caudal vena cava, and all over the body. The hepatic portal system and the liver regulate the amount of carbohydrates in the blood. A prepared microscope slide and a compound microscope are necessary to see these smaller vessels.

COMPARISONS OF VENOUS SYSTEMS IN THE CAT AND MAN (Figs. 4-1 to 4-4)

1. The external jugular vein in the cat is larger than the internal jugular; hence it is interpreted that the axillary and external jugular veins unite to form the brachiocephalic vein. However, in man the internal jugular is larger than the external jugular vein, and therefore it is interpreted that the subclavian and internal jugular veins unite to form the brachiocephalic vein.

2. Compare the diagrams of the caudal vena cava and its tributaries in the cat and man (Figs. 4-1 and 4-3).

3. The great saphenous is relatively larger and longer in man, passing up the thigh and

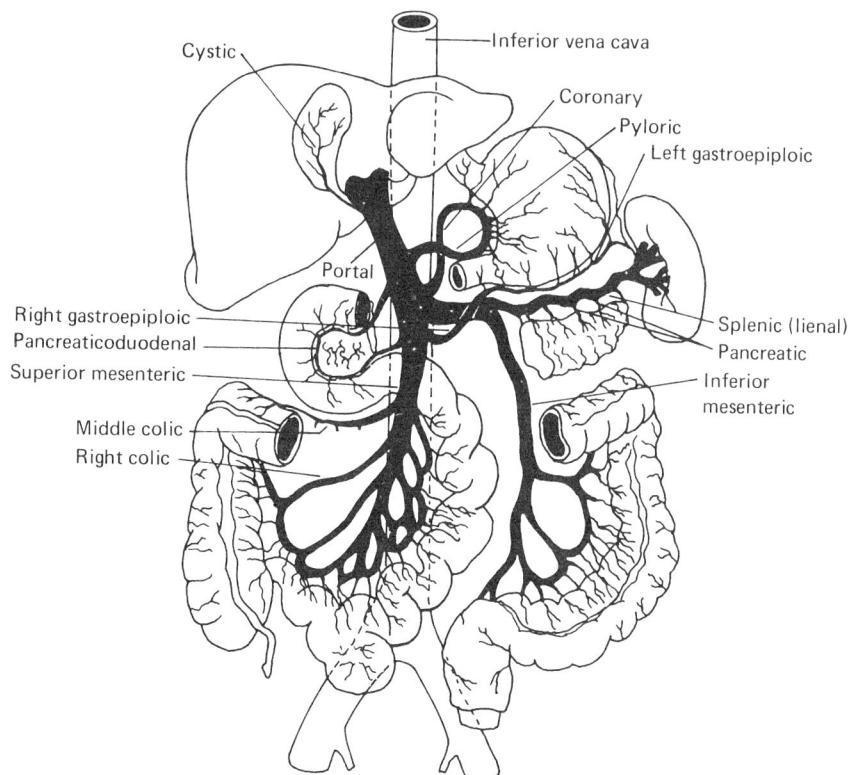

Fig. 4-4. Hepatic portal system of man. (Modified from Anthony, C. P., and Kolthoff, N. J.: Textbook of anatomy and physiology, St. Louis, 1975, The C. V. Mosby Co.)

joining the external iliac in the pelvic region; in the cat the medial saphenous joins the femoral a short distance above the knee.

4. The internal iliac and its branches are present but relatively less conspicuous in man.

5. The caudal vein is relatively large in the cat in comparison with its homologue in man, the middle sacral vein.

6. Two veins usually drain the right kidney of the cat. This condition occurs less often in man.

7. The thoracic duct enters the left external jugular vein in the cat but enters the left subclavian vein in man.

8. The thoracic duct in the cat has a relatively larger portion that is bifurcated or double.

9. In the cat the cephalic vein connects with the transverse scapular, whereas in man this vein enters directly into the axillary vein.

10. The saphenous vein of the thigh and leg is a major superficial vessel in man but is relatively much smaller in the cat.

11. In man a posterior external jugular vein drains the back of the head and neck and enters the external jugular vein on the lateral surface of the neck. The cat does not have this vein.

12. In man there is a hemiazygos (left vena azygos) vein on the left side of the chest; it is absent in the cat.

13. In man there is no intermediate vein connecting the popliteal vein with the lower end of the gluteal inferior (caudal gluteal) vein as in the cat.

14. Because man is in the erect position the blood in the veins of the leg and most of the body has to flow upward against gravity. Often there is not enough push, and it starts down the leg against the valves, producing the knotty enlarged condition known as **varicose veins.** There is no known record of cats having varicose veins.

Name _____

Date _____

REVIEW QUESTIONS ON VENOUS SYSTEM

1. Trace the blood from the popliteal vein to the left lung.

2. Make a small sketch and describe the anomalies you have observed in the study of veins.

3. What are the principal parts and the main function of the hepatic portal system?

4. How are the median, brachial, and axillary veins related? What term is listed in the nomenclature to include these veins?

5. What two veins unite at the first rib to form the brachiocephalic?

6. Name four tributaries of the brachiocephalic.

7. Why is the blood system a two-way system?

8. What are the two principal vessels entering the heart? Name the tributaries of each.

9. How are the popliteal, femoral, and external iliac veins related?

10. Describe the location of the right and left common iliac veins. How is each formed?

LYMPHATIC SYSTEM (Fig. 4-5)

The lymphatic system is a one-way system, since its lymph flows only toward the heart. Its capillaries are all over the body, adjacent to the skin, internal organs, brain, and bones, but are so small that they are seldom seen in ordinary dissection. Lymph is received from three minute sources: (1) from **arterial capillaries**, or **sinusoids**, of lymph nodes, spleen, thymus, and tonsils where it receives various white blood corpuscles and the plasma; (2) from the **small lymph sinuses** close to the various cells of the body, where it receives toxic or poisonous, broken-down by-products of the metabolic or physiological processes of the cytoplasm; and (3) from the **lymph capillaries** in the walls of the small intestine known as **lacteals**, which absorb the fats of digestion, called **chyle**. Fats filter through the mesenteric nodes, previously mentioned, which are the largest nodes in the cat. (They are largest in carnivorous animals. Several lymph nodes are relatively larger in the cat than in man.) The lymphatic system filters its lymph through the lymph nodes and returns these inactive substances to the venous system, which in turn takes them to the heart and hence to the lungs and kidneys where most unusable or injurious substances are removed from the blood and eventually eliminated from the body. It must be remembered that the lymph nodes are most effective because of the activities of their cells, which render many toxic substances harmless before they enter the bloodstream.

The red blood cells in the adult originate in the marrow of the long bones and pass directly into the veins.

The following structures are the principal parts of the lymphatic system.

The **thoracic duct** is shown in Fig. 4-5. It is thin walled, usually reddish brown in color, and about the size of the lead in a pencil. It is often "beaded" or irregular in size because of the dilations caused when the lymph flows backward against the numerous valves. The duct is often seen behind, or dorsal to, the peritoneum of the thoracic cavity, close against the dorsal aorta immediately above the diaphragm. With a dissecting needle carefully dissect off the peritoneum that covers it. The thoracic duct mostly lies ventral to the intercostal arteries but passes dorsal to them in the upper thoracic region. Dissect the thoracic duct caudal to the diaphragm where it dilates into the **receptaculum chyli**, or **cisterna chyli**. The cisterna chyli is slightly to the right of the aorta at the diaphragm, but the thoracic duct passes over the left side of the esophagus and trachea cranially to empty into the termination of the left external jugular vein. Caudal to the kidneys, small lymphatic tributaries may be seen uniting to form the thoracic duct. Trace the duct forward as it passes behind the left brachiocephalic vein and left subclavian artery. It soon makes a bend to the left and enters the left external jugular vein in the cat, while in man it opens into the **left subclavian vein**.

The lymph in the lymphatic system is believed to flow continuously, largely by the steady absorption of lymph from arterial and venous capillaries and sinuses, by movements of muscles, and probably by some pressure from the heartbeat through arteries and sinuses. If the venous system is well injected, some of the injecting fluid may have gotten past the valves and into the upper part of the thoracic duct.

There are many hundreds of lymph nodes. Most of them are quite small and are embedded mostly in the fascia and connective tissue. Major lymph node areas are called lymphocenters. In Fig. 4-5 the various nodes are located although not described separately in the text.

In man many fairly large lymph nodes are located in the groin, axilla of the arm, at the base of the lungs, about the heart, in the upper neck region, and about the mammary glands. **Cancer cells** are often found within these nodes, and infection spreads from one to another; hence, in many operations it is necessary to remove these centers of infection.

In the cat the **main lymphatic duct** from the **mesenteric lymphocenter** to the **cisterna chyli** often has a beaded appearance caused by lymph backing up against the valves. Small, uninjected lymph vessels are sometimes seen in the mesentery of the small intestines. Here they are called **lacteals**, since they **absorb** and **transport** chyle, which contains fat. As chyle or lymph filters through the nodes, lymphocytes are added to the lymph and thus find their way into the blood system.

The **second main trunk** is the **right lymphatic duct**, which drains the right side of the thorax, right forelimb, and the right side of the neck and head. Identify it as it enters the terminal part of the **right external jugular vein**. The **tonsils**, **thymus**, and **spleen** are known as

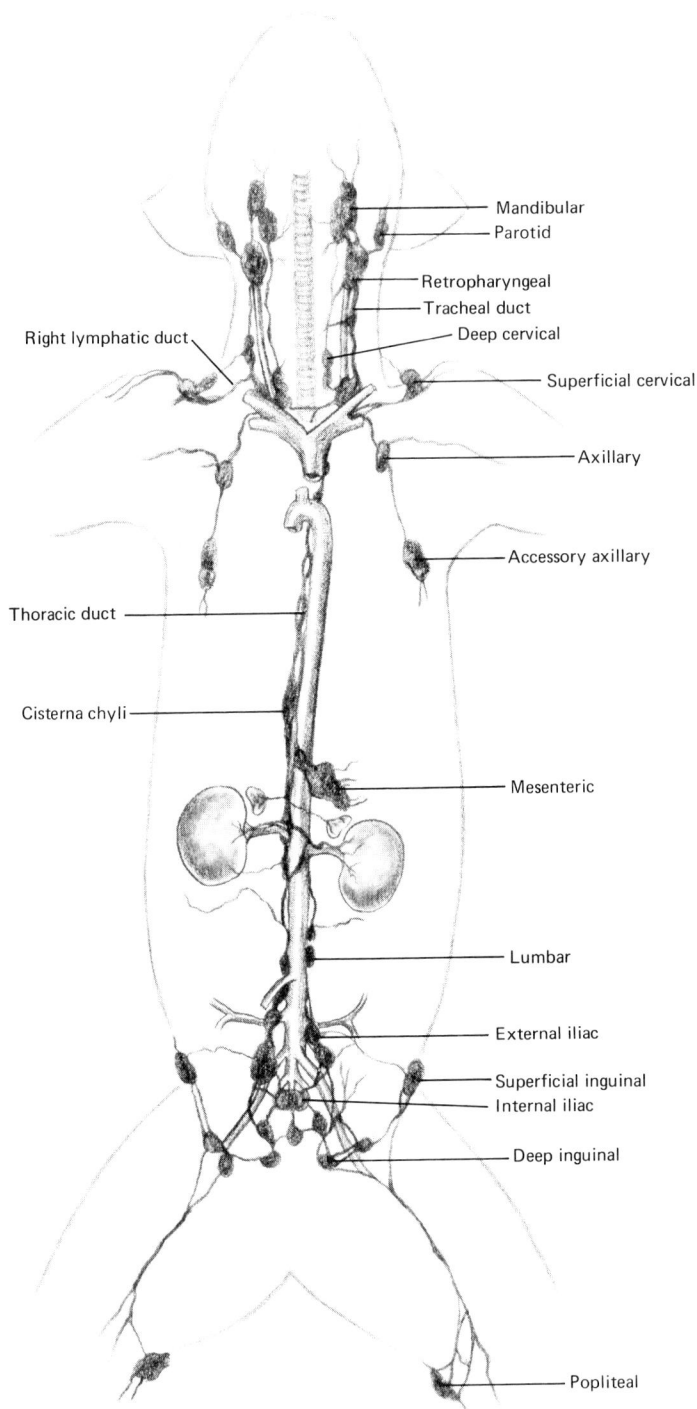

Fig. 4-5. Lymphatic system.

Labels:
- Mandibular
- Parotid
- Retropharyngeal
- Tracheal duct
- Deep cervical
- Right lymphatic duct
- Superficial cervical
- Axillary
- Accessory axillary
- Thoracic duct
- Cisterna chyli
- Mesenteric
- Lumbar
- External iliac
- Superficial inguinal
- Internal iliac
- Deep inguinal
- Popliteal

lymphoid organs, since they resemble lymph glands histologically and are considered modified lymph glands. The lymphatic system is sometimes referred to as a **salvaging system,** since it helps conserve many substances that can be purified and returned to the blood circulation for reuse.

REVIEW QUESTIONS ON LYMPHATIC SYSTEM

1. What are the names of the principal parts of the lymphatic system?

2. Where does the thoracic duct join the venous system in the cat and in man?

3. Explain how the lymphatic system is a one-way system. How is lymph able to flow more or less continuously? Where does lymph come from?

4. Where are the principal lymph nodes located?

5. Why are lymph vessels occasionally beaded?

6. Name and locate each of the lymphoid organs.

7. What blood corpuscles are known to originate in lymph nodes?

8. What peculiar cells enable the lymph nodes to function most effectively?

9. What is the general distribution and location of lymph capillaries?

10. Which organs resemble the lymph nodes and are called lymphoid organs?

11. What are the main contributions of the spleen and lymph glands to the blood?

12. What are phagocytes and where are they produced?

13. What is chyle?

FIVE
Arterial system and heart of the cat

In specimens properly prepared for dissection the arteries have been injected with red latex or a colored starch mass. In a former exercise, when the thoracic organs were identified, the fat and pericardium surrounding the heart were dissected away. Identify the **aortic arch,** which is the bend of the base of the **dorsal aorta,** to the left of the entrance of the **cranial vena cava** with the **azygos.** Note that the aortic arch curves to the left and passes caudally along the mid-dorsal line of the abdominal cavity. This aortic arch represents the fourth, or systemic, arch, which has persisted on the left side of the embryo. Find the branches given off in the following order.

ARTERIES WITH BASES WITHIN THORACIC CAVITY (Fig. 5-1)

1. The **coronary arteries** are usually injected with red latex and are found in the ventral and dorsal walls of the heart. They arise from the beginning of the **aorta** (Figs. 5-1 and 5-3). The coronary veins open into the coronary sinus, which opens into the caudal part of the right atrium. These parts usually are poorly injected because the injection is made against the valves.

2. The **pulmonary artery** extends from the right ventricle (Figs. 5-1 and 5-3) between the two auricles. This vessel divides ventral to the aortic arch for each lung. In order to expose the arteries better, cut the cranial vena cava in front of the heart and pull the brachiocephalic veins forward. As this is done, observe the vertebral veins passing dorsalward to enter the transverse foramina of the last cervical vertebra; they also pass through each of the transverse foramina of the cervical vertebrae up to the skull.

3. The **vertebral artery** is by the side of the vertebral vein, on the right side of the cat, as it enters the transverse foramen of the last cervical

vertebra. **Do not dissect** further. It is small and will probably be seen for only about a centimeter. The two vertebral arteries extend forward and unite along the ventral surface of the brain to form the **basilar artery** (see Fig. 7-5).

4. The **brachiocephalic artery** is the largest branch from the aortic arch. It gives off the **left common carotid artery** first and then the **right common carotid** and continues as right subclavian.

5. The **right subclavian artery** continues through the thoracic inlet. After turning around the first rib it becomes the axillary artery.

6. The **left subclavian artery** arises next from the aorta and turns around the left first rib.

7. The **internal thoracic artery** of each side arises from the subclavian artery close to the first rib and courses caudally on the dorsal surface of the sternum deep to the transversus thoracis muscle. At the diaphragm it divides into **cranial epigastric** and **musculophrenic** branches. The cranial superficial epigastric artery anastomoses with the caudal superficial epigastric in the mammary gland or comparable area in the male.

8. The **intercostal arteries,** except the first few, may be seen branching from the aorta.

9. The **costocervical artery** arises from the subclavian just cranial to the vertebral and sends a branch to the dorsal thoracic wall and the supreme intercostal artery, from which come the first few intercostal arteries.

10. The **superficial cervical artery** arises from the subclavian as it nears the first rib. It goes through the thoracic inlet to the neck and shoulder.

11. The **bronchoesophageal artery** is usually single and arises from the right side of the aorta at the base of the heart or from an intercostal artery on the right. It divides into two

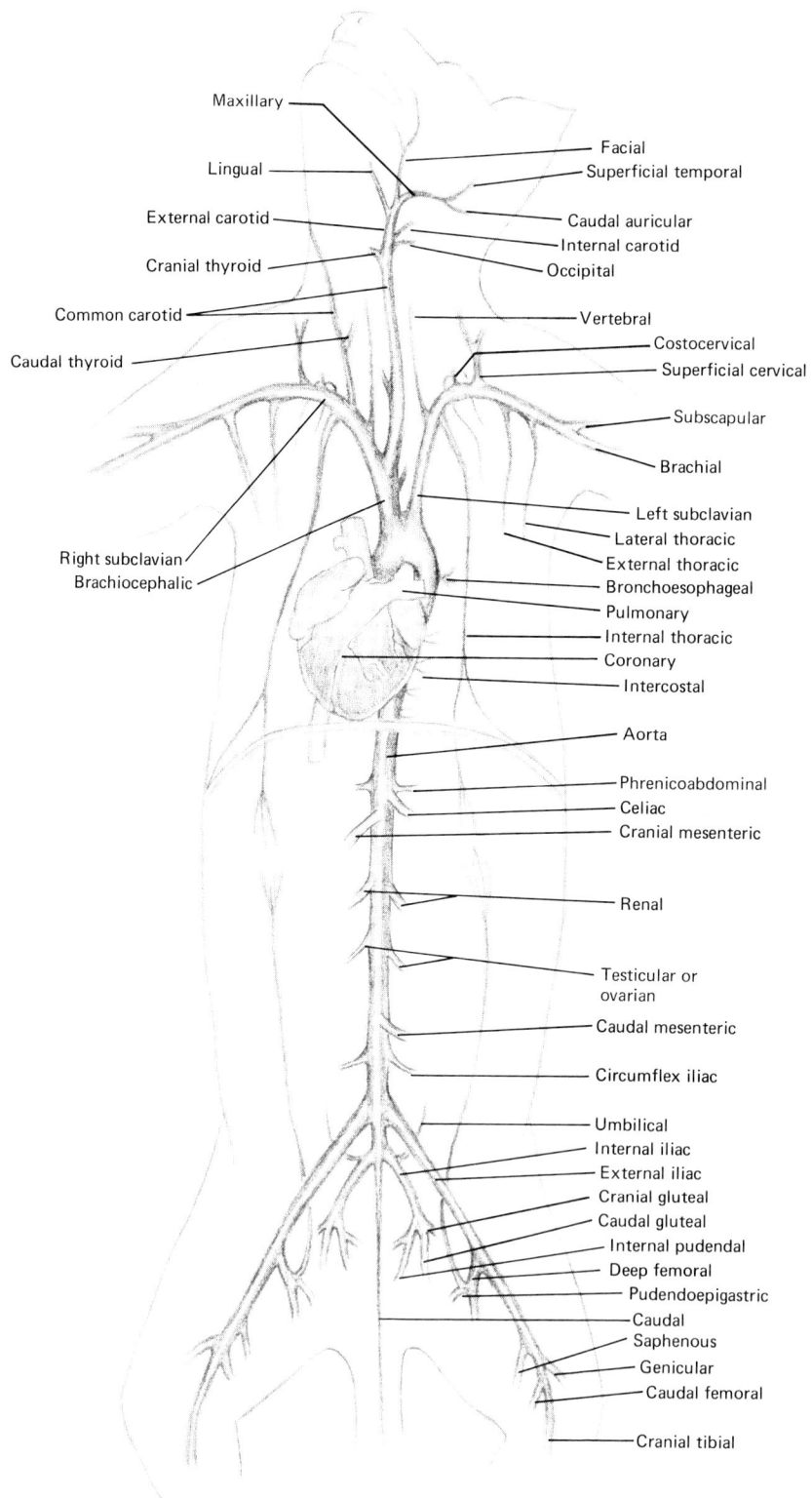

Fig. 5-1. Arterial system.

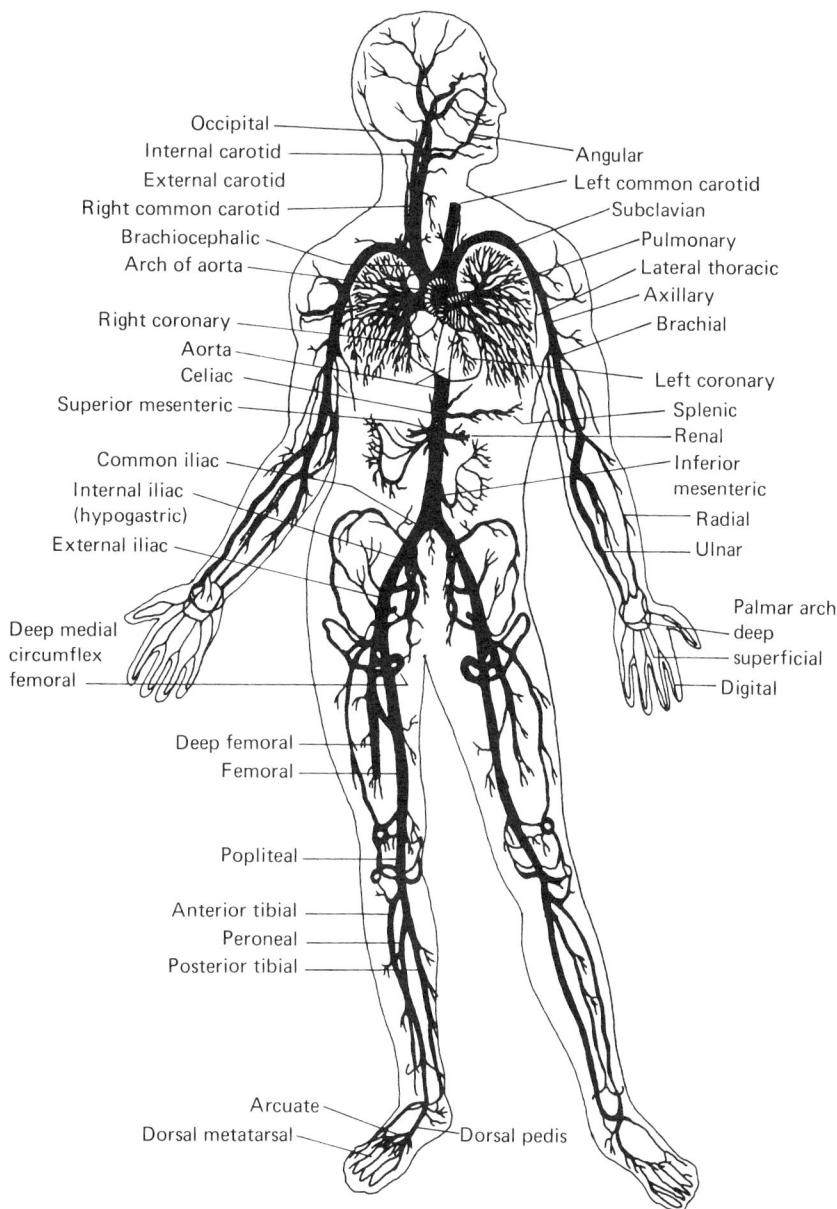

Fig. 5-2. Human arterial system. (Modified from Anthony, C. P., and Kolthoff, N. J.: Textbook of anatomy and physiology, St. Louis, 1975, The C. V. Mosby Co.)

bronchial arteries and an esophageal artery. This is a nutritional circulation to the lungs.

ARTERIES TO UPPER PART OF NECK AND HEAD (Fig. 5-1)

The branches of the **common carotid artery** lie medial to the salivary glands. Therefore, turn to Fig. 2-10 where the salivary glands and adjacent veins are shown. After they are identified, return to this page. Dissect the following

arteries so that they can be demonstrated to the instructor.

1. Start with the **left common carotid artery** where it branches from the brachiocephalic artery. Dissect away the muscles and connective tissue, and expose it and its branches up into the head. The name of each branch is determined by the part supplied and not by its position of origin. (a) The **caudal thyroid artery** is small and extends dorsally to supply the

123

serratus cervicis. (b) The **cranial thyroid artery** is small and extends medially near the larynx. It supplies the thyroid gland and sternothyroideus and sternohyoideus muscles. (c) The **cranial laryngeal artery** (not shown) branches off about a quarter of an inch cranial to the cranial thyroid artery and courses medially to the muscles of the larynx. (d) The **occipital artery** has its origin almost opposite that of the cranial laryngeal artery. It is long and extends forward to the deep muscles of the neck and the occipital region of the skull.

2. The **common carotid artery** divides at the base of the skull into **external carotid** and **internal carotid arteries.** (a) The **lingual artery** is fairly large as it branches from the external carotid near the bend and enters near the center of the extreme caudal portion of the tongue. The lingual artery usually extends some distance nearly parallel with the hypoglossal nerve, which also courses forward to enter the tongue. (b) The **facial artery** branches from the external carotid and passes beneath the digasticus muscle toward the corner of the mouth, where it divides into several branches, the most important being those to the lips. (c) The **caudal auricular artery** branches from the external carotid and goes caudal to the external acoustic meatus. (d) The **superficial temporal artery** branches off between the ear and the corner of the mouth to supply the region between the base of the ear and the eye. (e) The **maxillary artery** is a continuation of the external carotid and supplies the maxillary region.

Dissect the **internal carotid artery** slightly caudal to the origin of the **lingual artery** and among the deep muscles. In Fig. 7-5 the internal carotid is shown entering the arterial circle from each side. The arteries supplying the ventral surface of the brain will be considered after the brain has been removed from the skull and a discussion of the structure has been given.

ARTERIES TO LOWER PART OF NECK AND LIMBS (Fig. 5-1)

Return to the **right subclavian artery** as it leaves the **brachiocephalic artery.** Within the thoracic cavity the name subclavian is used; just outside the thoracic wall it becomes the axillary artery. In the upper arm, it is the brachial, and then in the forearm, it is the median.

1. The **external thoracic artery** branches from the **axillary artery** and courses to the middle of the pectoral muscles and to the latissimus dorsi. The **lateral thoracic artery** also goes to the pectorals.

2. The **subscapular artery** branches from the axillary artery where it continues as brachial. This is in the area between the subscapularis and teres major. It gives off the caudal circumflex humeral and thoracodorsal branches.

3. The **brachial artery** gives off the cranial circumflex humeral and deep brachial branches to the arm.

ARTERIES OF ABDOMEN AND LIMBS (Fig. 5-1)

1. The **celiac artery** arises from the dorsal aorta immediately caudal to the diaphragm and soon divides into three main branches. (a) The **hepatic artery** gives off the gastroduodenal artery to supply the pyloric portion of the stomach and the pancreaticoduodenal artery to supply the pancreas and duodenum. The main hepatic artery passes to the liver parallel to the common hepatic duct and the portal vein. (b) The **left gastric artery** supplies the lesser curvature of the stomach and part of the ventral surface, which is its embryonic left side. (c) The **gastrosplenic (splenic) artery** is the largest of the three and supplies the dorsal side of the spleen and a portion of the greater omentum.

2. The **cranial mesenteric artery** arises close behind the base of the celiac artery, passes between the celiac and cranial mesenteric ganglia, and gives off the pancreaticoduodenal artery, which, as mentioned, supplies the pancreas and duodenum. The large middle colic artery supplies the transverse colon, and the ileocolic (ileocecocolic) artery supplies the lower part of the ileum, cecum, and ascending colon. The main artery continues to the jejunum.

3. The **parietal arteries** arise segmentally to supply the body wall on each side of the dorsal aorta.

4. The **phrenicoabdominal artery** arises from each side of the aorta near the crus of the diaphragm. One branch courses to the suprarenal, whereas the main trunk goes to the abdominal wall. The phrenic portion supplies the crus of the diaphragm.

5. A **renal artery** supplies each kidney, but the right one is a bit cranial to the left one. Sometimes the artery branches before entering the kidney.

6. The **genital (testicular or ovarian) artery** is slender and arises caudal to the renal artery. If you have a male, trace the testicular artery to where it leaves the body through the inguinal canal as part of the spermatic cord. The ovarian artery is much shorter embryologically; the testis differentiates close to the kidney where its blood supply is established. It migrates through the inguinal canal, taking the blood vessels and nerves with it. Usually the ovary migrates comparatively little.

7. The **single caudal mesenteric artery** courses to the descending colon and rectum.

The dorsal aorta continues to the pelvic region and gives off the two external iliac arteries, which continue into the pelvic limb as femoral arteries. The internal iliac arteries arise caudal to but not with the external iliac arteries. In the midline from the aorta's termination is the caudal, or middle sacral, artery.

8. The **external iliac artery** gives off the following branches. (a) The **circumflex iliac artery** arises immediately and passes laterally into the body wall. (b) The **deep femoral artery** branches caudally near the pubis and courses to the medial muscles of the thigh. Near its origin it gives off the pudendoepigastric trunk, which divides into external pudendal through the inguinal canal and the caudal deep epigastric (caudal abdominal) to the lateral edge of the rectus abdominis.

9. The external iliac artery enters the thigh as the **femoral artery,** which gives off the following branches. (a) The **saphenous artery** is a substantial branch that arises medially and runs subcutaneously from the distal end of the

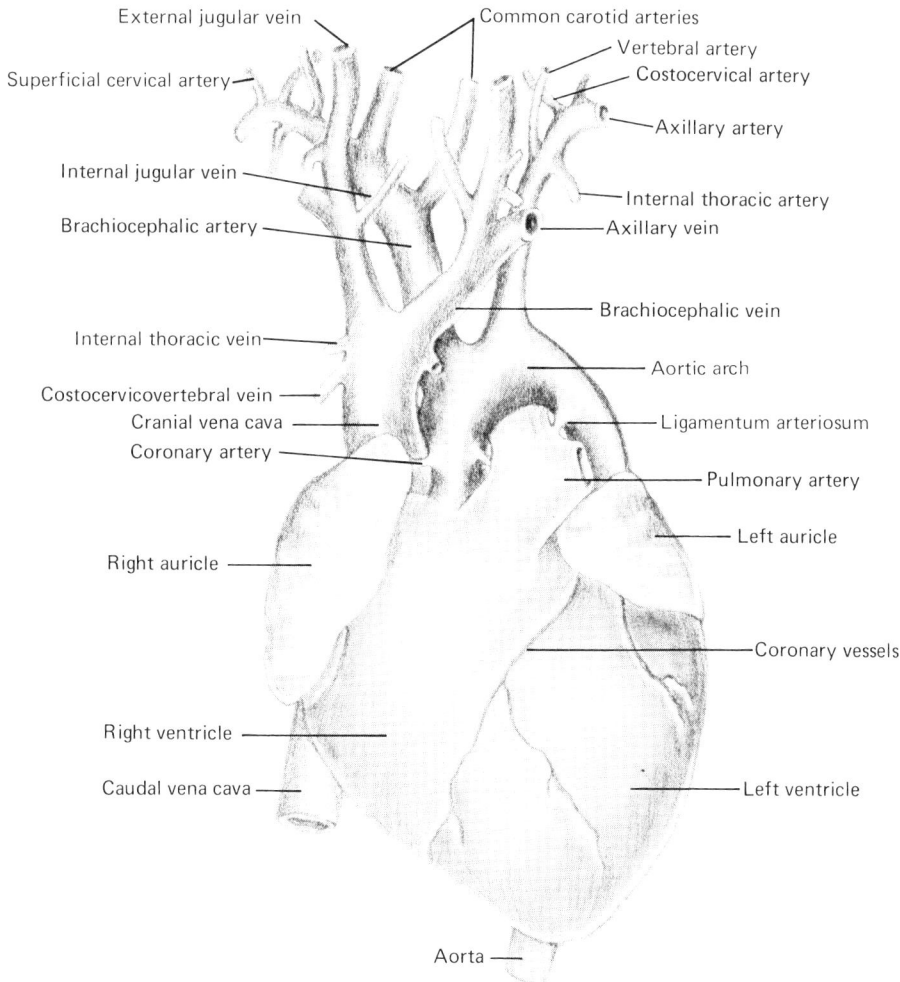

Fig. 5-3. Heart, left ventral view.

External jugular vein
Superficial cervical artery
Internal jugular vein
Brachiocephalic artery
Internal thoracic vein
Costocervicovertebral vein
Cranial vena cava
Coronary artery
Right auricle
Right ventricle
Caudal vena cava

Common carotid arteries
Vertebral artery
Costocervical artery
Axillary artery
Internal thoracic artery
Axillary vein
Brachiocephalic vein
Aortic arch
Ligamentum arteriosum
Pulmonary artery
Left auricle
Coronary vessels
Left ventricle

Aorta

femoral triangle. On the medial side of the leg it divides into cranial and caudal arteries. (b) The **genicular artery** passes toward the patella. (c) The **caudal femoral artery** arises caudally at the origin of the gastrocnemius.

10. The popliteal artery becomes the **cranial tibial artery**, which goes through the interosseus space.

11. The **internal iliac artery** gives off the following branches. (a) The **umbilical artery** is reduced to the form of a ligament but may supply blood to the urinary bladder. It lies in the lateral ligament of the bladder. (b) The **cranial gluteal artery** courses over the greater ischiatic notch to the gluteal muscles. (c) The **caudal gluteal artery** passes through the lesser ischiatic notch to the caudal muscles of the hip and thigh. (d) The **internal pudendal artery** goes to the structures of the ischial arch.

RELATIONSHIPS OF HEART (Fig. 5-3)

It is helpful to remove the heart from the body in order to dissect it properly. To do this the pericardium must be removed, and the lungs must be cut off close to their bases. Identify each of the following structures.

1. The **cranial vena cava** is formed by the union of the two **brachiocephalic veins**. Cut the brachiocephalic veins in two. Lift up the cranial vena cava and identify the **costocervical** and **internal thoracic veins**. Cut each off close to its base. The **azygos vein** (see Fig. 4-1), which enters the cranial vena cava close to the heart, is to be cut also.

2. Cut the **caudal vena cava** cranial to the diaphragm.

3. Cut the **right subclavian artery** (Fig. 5-4) halfway between its base and where it passes out of the thorax.

4. Cut the **left** and **right common carotid arteries**, which branch from the **brachiocephalic artery**.

5. Cut the **left subclavian artery** (Fig. 5-4) and the small tributaries that enter its base.

6. Cut the **dorsal aorta** near the heart and also cut the pulmonary arteries and veins.

7. Lift up the vessels that have been cut cranial to the heart and carefully pull the heart and its attached vessels away from the lower part of the trachea, bronchi, and base of the lungs, leaving them in place. Now the heart can be examined on all sides.

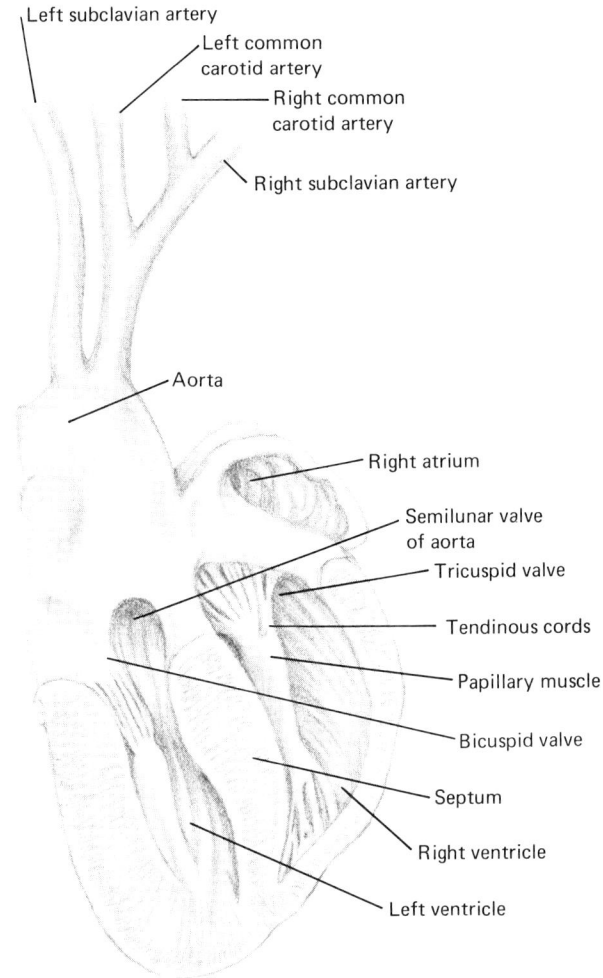

Fig. 5-4. Heart, opened.

DISSECTION OF HEART (Figs. 5-3 and 5-4)

This dissection will be done in the same order and direction as the flow of the blood.

First step: With the scissors clip a small hole in the **cranial vena cava** and in the **caudal vena cava**, close to their place of entrance into the heart. Insert a probe in each of the clipped openings and push them into the **right atrium**, where they will touch one another. Cut the ventral walls of the blood vessels and the right atrium with the scissors, following close to the probes. Remove the injecting material.

Second step: Place the end of a probe in the **right atrium** and push it down into the **right ventricle**. Now cut the ventral wall of the latter open by cutting down along the side of the probe to the lower part of the right ventricle. Remove the latex or other injecting material from the right ventricle and see the fleshy columns, or **trabeculae carneae,** on the inside of the wall. The blood leaves the **right ventricle** past the **bicuspid** and **semilunar valves** through the **pulmonary artery** on its way to the lungs. This blood vessel extends upward from the **right ventricle** close to the ventral surface of the heart and between the two auricles, where it divides into **right** and **left pulmonary arteries.**

Third step: By exercising some care, you can push the probe into the **pulmonary artery** from the inside of the **right ventricle.** When this has been done, cut with the scissors, following the probe as a guide, up to a point where the pulmonary artery divides to carry impure blood to each lung. Now the **tricuspid valves** between the right atrium and right ventricle can be seen better. There are three **cusps,** or **flaps,** and each is supported by fine fibers, or **chordae tendineae,** and **one papillary muscle.** The **semilunar valves** can also be seen between the right ventricle and the common pulmonary artery. There are two earlike projections, or **auricles,** one on the ventral wall of each atrium. These are usually somewhat more wrinkled and darker than the other parts of the heart. The **pulmonary artery** extends between the right and left auricles and divides, sending a blood vessel to each lung. The small **ligamentum arteriosum** extends from the base of the **left pulmonary artery** to the **aortic arch** of the dorsal aorta. This ligament is the remnant of the distal end of the left embryonic **sixth aortic arch,** which closes following birth, thus forcing all blood from the right ventricle to go to the lungs to be oxygenated. Add colored arrows to indicate purity and direction of flow in these parts.

Fourth step: Cut transversely with the scissors through the ventral wall of the left atrium and extend the cut through the **left auricle.** Remove any injecting material.

Fifth step: Insert a probe into the left atrium, push it down into the left ventricle, and extend the cut past the bicuspid valves with the scissors down to the very tip end of the cavity. Dig and wash out the contents of the left ventricle and the left atrium and dry with paper towels. Now you should be able to probe from the inside of the left atrium and find the openings of the right and left **pulmonary veins** (not shown), which bring oxygenated blood from the lungs into the left atrium.

Between the left atrium and left ventricle you may now find the **bicuspid valves.** Observe the relative thickness of the walls of each of the four cavities of the heart.

Sixth step: If the base of the dorsal aorta has been well injected with latex, the vessels may be split open with the scissors and the latex removed intact. The latex often pushes down against the **semilunar valves,** leaving a good impression of them and of the bases of the **coronary arteries** immediately above. After the base of the aorta has been split above these valves, you can push a probe past the valves into the left ventricle. The base of the dorsal aorta is the **fourth,** or **systemic, aortic arch** of the left side.

Place the end of your thumb in the **right atrium** and the forefinger of the same hand in the **left atrium.** Now try to rub the ends of the thumb and finger together and thus find the thinnest place in the septum separating the two atria. Hold this up between your eyes and the light, and you will see a very thin spot. This is the **fossa ovalis** and represents the place where the opening between the two atria existed during embryonic development. This opening was called the **foramen ovale,** and it closed soon after birth. Sometimes in the human baby this opening does not close. Under these conditions all the blood does not go into the right ventricle and to the lungs to be purified, but some passes through this opening; hence the baby does not have his blood sufficiently purified and is called a "blue baby" because of his dark or bluish color. This condition may also result if the ductus arteriosus fails to close.

CORONARY CIRCULATION

There are two kinds of circulation: (1) **functional,** which is the flow of blood through the atria and ventricles, as just described; and (2) **nutritional,** which is the circulation of the blood through the coronary arteries and veins to supply the walls of the heart. The coronary arteries branch from the deep base of the aortic arch. These are probably injected red near the surface of the heart walls. The **coronary sinus,** into which the **coronary veins** from the wall of the heart empty, is covered with fat on the caudal surface of the **right atrium** ventral to the caudal vena cava.

STAGES OF DEVELOPMENT

The embryological development of the cat and human hearts shows recapitulation quite nicely. In brief, the heart passes through the following stages: at first it is quite similar to an elasmobranch fish, having **one atrium** and **one ventricle;** later it has **two atria** and **one ventricle,** as in a frog; and still later the ventricle divides and there are **two atria** and **two ventricles,** as in a bird or mammal.

In the typical vertebrate animal, such as the dogfish shark *(Squalus acanthias),* or in the **embryos** of the cat or man, there are structures representing **six pairs of aortic arches.** Their names and numbers are as follows: first aortic arch, mandibular; second aortic arch, hyoid; third aortic arch, common carotid; fourth aortic arch, systemic; fifth aortic arch, innominate; and sixth aortic arch, the pulmonary. In the **adult** of cat and man those remaining are the **third** (common carotid) on both sides; the **fourth,** only on the left side, which is the base of the dorsal aorta, or the aortic arch; the **sixth,** with the proximal parts only, which are the bases of the pulmonary artery on each side. These remaining aortic arches have migrated from the pharyngeal region of the young embryo down into the thoracic cavity. All other parts of the six pairs of aortic arches of the embryo have disappeared in the adult cat and man.

When some of the structures of a higher type animal pass through embryonic stages similar to those of adults of lower members of the same group, this phenomenon is called recapitulation. There are many such occurrences in the embryonic development of mammals. The branchial arches, the aortic arches, the pharyngeal pouches, the heart, the ear, and the kidneys are a few examples that are interpreted as showing recapitulation. Recapitulation supports the theory of evolution.

SOME DIFFERENCES IN ARTERIES OF THE CAT AND MAN

1. The left common carotid artery branches from the aorta in man, but in the cat it branches from the brachiocephalic artery.

2. The internal and external iliac arteries arise separately from the dorsal aorta in the cat, whereas there is a common iliac artery in man from which both arise.

3. The dorsal aorta passes caudally in the cat and divides at the first sacral vertebra, whereas in man it divides at the fourth lumbar vertebra.

4. The internal carotid arteries are large in man, whereas they are relatively small in the cat.

5. In man the superior and inferior phrenic arteries supply the diaphragm, whereas in the cat there are only the phrenicoabdominal arteries.

6. The deep femoral artery is relatively larger and more extensively developed in man than it is in the cat.

7. The erect position of man puts a much greater strain on the heart to pump the blood because of the pull of gravity.

REVIEW QUESTIONS ON ARTERIAL SYSTEM

1. How many pairs of aortic arches are represented in a typical vertebrate (embryo or adult) animal? Which is retained from the embryo as the base of the aortic arch?

2. Give the names and numbers of the aortic arches that are present in an adult cat or in man.

3. State three places where blood or the plasma of blood is purified.

4. Trace the blood as it passes from the left hind limb of a cat until it reaches the right ear and then returns to the right front limb.

5. Trace carbon dioxide from the inferior mesenteric vein of the cat until it is eliminated from the body.

6. Trace nitrogenous waste from the time it enters the right brachiocephalic vein until it is eliminated from the body.

7. What is the significance of the fossa ovalis of the heart and how did it come about?

8. What is the "pulse" of man, and where can it be demonstrated? (Use your own knowledge to answer.)

9. What is the significance of the ligamentum anteriosum?

10. Name and locate the bases of the blood vessels that supply nutriment to the heart.

11. Where do most red blood cells (erythrocytes) originate in the adult?

12. Name and locate three sets of valves in the heart.

13. What are the names of the continuations of the subclavian artery in the different regions of the forelimb of the cat?

14. Trace the circulation of the blood from the mesenteric lymph node to the left ear and back to the hepatic artery.

15. Trace the blood from the saphenous vein to the forearm and back to the hypogastric artery.

16. Explain the functional and the nutritional circulations of the blood.

SIX

Regions of the head of the cat

In this exercise two students should work together, bisecting the head, pharynx, and part of the neck of one cat and leaving the other cat skull to be dissected in such a way as to remove its brain intact and permit a dorsal and a ventral view examination of the entire brain.

The pharynx in a vertebrate animal is that part of the alimentary canal into which gill clefts, or gill slits, open or from which pharyngeal pouches extend in the embryo. In order to examine the pharynx of the cat and its relationships, bisect the entire skull and neck down to the upper parts of the trachea and esophagus by making a median sagittal cut. You can probably do this most easily with a thin, sharp, fine-toothed saw. However, before using the saw, cut the softer parts with scissors and a sharp knife. Begin by cutting open the upper part of the trachea and larynx along its median ventral line. Continue this cut forward with the knife, cutting the base and entire tongue in half along its median sagittal plane.

Use the saw, beginning between the nasal bones and back between the frontal bones on the dorsal surface. As you saw deeper, be certain your saw strikes the cut made with scissors and knife on the ventral surface. The skull and neck can then be divided completely along the median sagittal plane with little difficulty if care and some skill are exercised in keeping the saw in the exact median plane. Identify the following structures.

1. The **mouth,** or **oral cavity,** develops by an infolding of the skin and is thus lined with epithelial ectoderm. The roof of the mouth is lined with thick transverse ridges of mucosal epithelium known as rugae, which assist in swallowing as the dorsal surface of the tongue contacts them. The tongue is covered with several kinds of cornified papillae. The teeth are believed to be homologous with placoid scales.

2. The **tongue** forms a part of the floor of the mouth and pharynx. It arises beneath the pharynx and pushes forward into the mouth during embryonic development. Several muscles contribute to its formation. On the base of the tongue is a small pit, the **foramen cecum,** which marks the place where the **thyroid gland** passed in its embryonic migration to its position in the adult, caudal to the laryngeal cartilages.

3. The **nostrils** and **nasal cavity,** like the mouth, are lined with ectoderm, since they push in from the outside and join the pharynx by the internal apertures, or **choanae.** The incisive, maxillary, and palatine bones form the **hard palate,** which separates the mouth and nasal cavities. The **turbinate bones** push into each nasal cavity from the lateral side of each nasal chamber and are parts of the ethmoid, nasal, and maxillary bones. The **soft palate** is caudal to the **hard palate,** and in man there is a fingerlike process that hangs down from its center known as the **uvula** (Fig. 6-2). This is absent in the cat.

4. Two air cells, or air sinuses (paranasal sinuses), are above the **cribriform plate** and nasal chambers. The more rostral one is the **ethmoid sinus,** and the cranial one is the **frontal sinus.** These communicate with the nasal chamber in mammals, but the erect position of man's body makes drainage of the sinuses difficult and is one of the factors involved in sinus trouble. Another air space is the **sphenoid sinus** within the sphenoid bone rostral to the sella turcica. Above the hard palate in the maxilla is the **maxillary sinus** (not shown).

5. The cranial cavity contains the brain. The **cerebrum** and **cerebellum** are partially

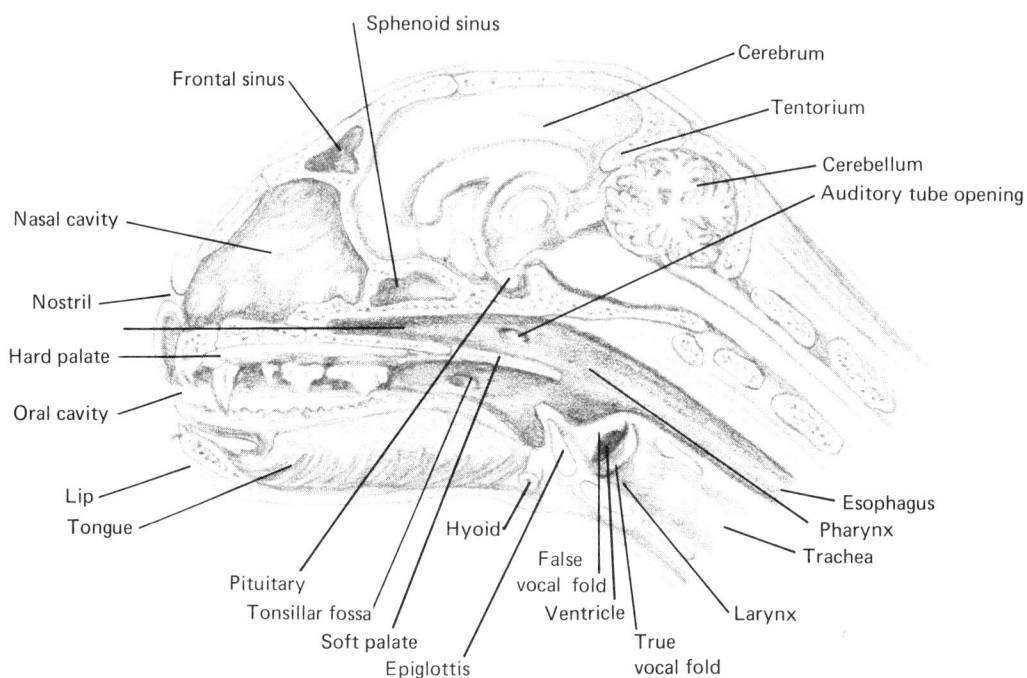

Fig. 6-1. Head of the cat, median sagittal view.

Sphenoid sinus

Cerebrum

Frontal sinus

Tentorium

Cerebellum

Auditory tube opening

Nasal cavity

Nostril

Hard palate

Oral cavity

Lip

Tongue

Esophagus

Pharynx

Trachea

Hyoid

False vocal fold

Larynx

Pituitary

Tonsillar fossa

Ventricle

Soft palate

True vocal fold

Epiglottis

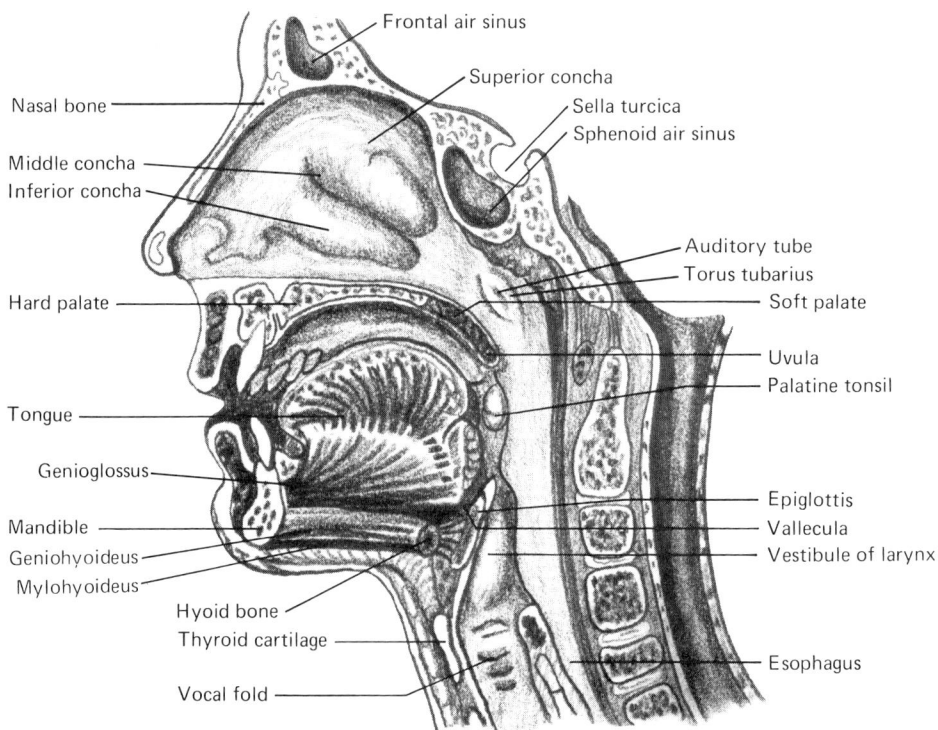

Fig. 6-2. Human mouth, larynx, pharynx, and nasal cavity, median sagittal view.

Frontal air sinus

Superior concha

Sella turcica

Sphenoid air sinus

Nasal bone

Middle concha

Inferior concha

Auditory tube

Torus tubarius

Soft palate

Hard palate

Uvula

Palatine tonsil

Tongue

Genioglossus

Epiglottis

Mandible

Vallecula

Geniohyoideus

Vestibule of larynx

Mylohyoideus

Hyoid bone

Thyroid cartilage

Esophagus

Vocal fold

separated by the **tentorium** in the cat. This is an ossified portion of the dura mater that forms extensions on the parietal bones. It is absent in man. In the cranial floor is a depression, the sella turcica, that contains the **pituitary gland.** The membranes covering the brain are collectively known as the **meninges:** the outer is the dura mater; the middle is the arachnoid; and the inner is the pia mater (see Fig. 8-5).

6. The **eustachian tube** is a passageway from the **pharynx** to the vestibule of the middle ear. It represents the vestigial remains of the inner portion of the first gill cleft. The **opening of the eustachian tube** into the pharynx is just caudal to the **hamular process** of the **sphenoid bone.** The opening appears to be in the lateral wall of the nasopharynx, because in the cat the soft palate grows caudally and divides the rostral part of the pharynx into **nasal** and **oral portions.**

7. The **palatine tonsil** is located in a small depression, the **tonsillar fossa,** in the lateral wall of the **oral pharynx** and is about the size of a grain of rice. Part of it represents the endodermal lining of the second **pharyngeal pouch,** or gill cleft. The palatine tonsil is removed in a **tonsillectomy** in man. Two other kinds of tonsils, which are not apparent in the cat, are the **pharyngeal** (**adenoids**) and the **lingual.** The former are in the middorsal portion of the pharynx at the base of the skull; and in man, when the pharyngeal tonsils enlarge, the internal nares are constricted or closed. The lingual tonsils are embedded in the lateral edges of the caudal part of the tongue.

8. The **larynx** (Fig. 6-1) lies ventral to the caudal portion of the pharynx; the cavities of the two are connected by the aditus laryngis. Projecting forward from the ventral portion of the larynx is the **epiglottis.** When the tongue is retracted in swallowing, it pushes the epiglottis dorsally over the aditus laryngis. The caudal cartilage forming a ring at the rostral end of the trachea is the **cricoid cartilage.** A wing that lies on each side rostral to it is the **thyroid cartilage,** or **lamina.** The wings join ventrally to form a median prominence, the Adam's apple of man. Medial to the thyroid cartilage on each side is the **arytenoid cartilage.**

Find the notch on the upper edge of the thyroid cartilage of your larynx. Push the end of your finger up and feel the **hyoid cartilage,** or bone. Between the body of the thyroid cartilage and the cricoid cartilage is the space filled in by the **cricothyroid membrane.** The hyoid bones consist of several sections embedded in the lateral wall and floor of the region immediately cranial to the larynx. They are anchored to the larynx and the base of the tongue. Dorsally they fasten to the area of the bulla tympanica. When all parts of the hyoid apparatus are present, as in the cat, there is a U-shaped arrangement of bones consisting of basi-, thyro-, cerato-, epi-, stylo-, and tympanohyoids.

9. Within the larynx are membranous folds, the **vocal folds** (**cords**), which extend along the sides of the lumen of the larynx from the arytenoid cartilage to the body of the thyroid cartilage. The **false vocal folds** are thickenings closer to the aditus and rostral to the **true vocal folds.** Between the true and false vocal folds is an opening, the **ventricle.** Between the vocal folds, right and left is the **glottis.** Adduction of the vocal folds and arytenoid cartilages closes the glottis.

10. Examine the **trachea, bronchi,** and **bronchioli.** Cut the tracheal tube open longitudinally and examine its cartilaginous rings. How do the bronchi and bronchioli differ? The **esophagus** lies between the trachea and the lower cervical vertebrae. Compare Figs. 6-1 and 6-2. What are the principal differences and similarities between the cat and man?

SOME DIFFERENCES IN THE CAT AND MAN

1. The soft palate of the cat divides the rostral part of the pharynx into the nasal and oral portions. This does not occur in man.

2. The fingerlike projection on the edge of the soft palate in man is the uvula. This is absent in the cat.

3. The salivary glands are relatively much larger in the cat than in man.

4. The submandibular gland in the cat is closely associated with the parotid, whereas in man it lies below the mandible.

5. The sublingual gland in the cat usually lies against the submandibular, whereas in man it lies beneath the floor of the mouth under the tongue only.

6. The palatine tonsil seems to be the only tonsil that is well developed in the cat, whereas man has three pairs of fairly large tonsils, the palatine, pharyngeal, and lingual tonsils.

7. The hyoid bone in the cat is composed of a number of distinct segments: the basi-, cerato-,

epi-, stylo-, and tympano- hyoids. These segments correspond to the lesser cornu of man, whereas the thyrohyoid of the cat is homologous with the entire greater cornu of man.

8. The lymph nodes adjacent to the submandibular and sublingual salivary glands are relatively much larger in the cat than in man.

9. The primary cause of the differences in the relationships of the skull and in the face of cat and man results from the enlargement of the cerebrum in man, which has pushed the skull dorsalward and the mouth, nose, and nasal chamber anteriorward or ventrally.

10. The size of the nose and nasal chamber has been relatively reduced in size in man, which is one of the factors causing a reduction in the keenness of smell.

11. The pharynx is normally held in a vertical position in man, whereas the pharynx of the cat is held at about a forty-five–degree angle or often in a horizontal position. It is generally recognized that the sinuses drain normally when the head is held in a horizontal position; therefore, man is much more likely to have sinus trouble than a cat. Expressed in another way and supported by facts of embryology, the head of man appears to be much more distorted than that of the cat, largely because of the increase in the size of the cerebrum.

REVIEW QUESTIONS ON UPPER DIGESTIVE AND RESPIRATORY REGIONS

1. Name the openings that lead into or away from the pharynx.

2. Name the different structures, in order, through which food passes in reaching the pyloric cecum.

3. Name four cartilages that form the larynx.

4. Name the three salivary glands and their respective ducts. (See Fig. 2-10)

5. Name the larger arteries and veins that lie close to the salivary glands.

6. Name the three kinds of tonsils and state their locations. Which are well developed in the cat?

7. Where is the thyroid gland located in reference to the cartilages of the larynx?

8. When a person swallows, what is the relationship between the passage through which the food goes and the larynx?

9. What is the location and significance of the cribriform plate?

10. Give the technical name and the principal structure of the "Adam's apple."

11. Name the structures through which air passes from the nostrils to the lungs in the cat and in man. What would you say are the principal differences in their respiratory tracts?

12. Compare the air (paranasal) sinuses in the cat and in man.

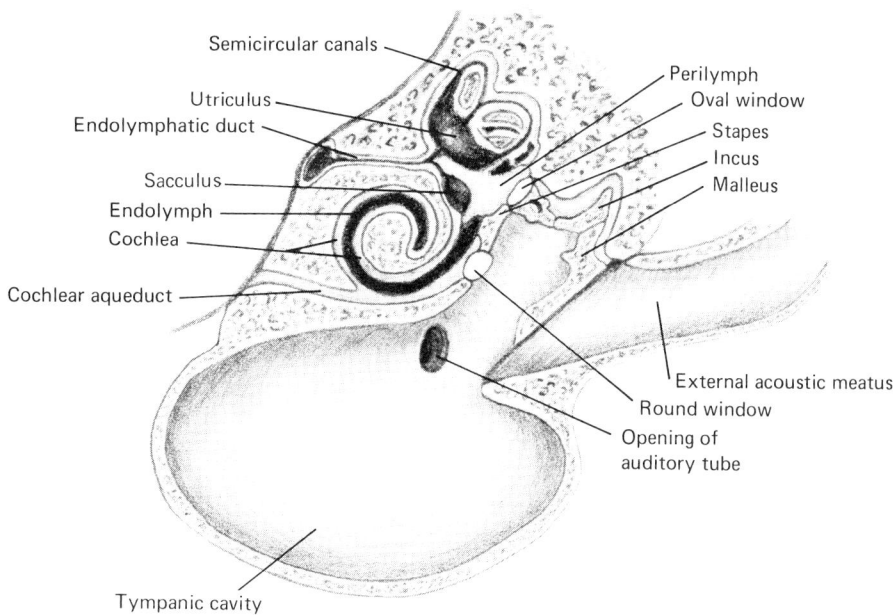

Fig. 6-3. Middle and inner ear.

DISSECTION OF EAR (Fig. 6-3)

Remove the skin from the entire skull, if this has not been done. The two functions of the ear are **equilibrium** and **hearing**. There are three principal parts of the ear of the cat: external, middle, and inner. Each is different in embryological origin. Each of these parts and its subdivisions will now be discussed.

1. The **external ear** consists of the **pinna**, or **concha**, which projects from the surface of the head. It is composed of elastic cartilage derived embryologically from the first two **visceral arches** and is covered with skin. The ear of the cat is pointed at the end. If you feel the edge of the upper cartilaginous portion of your own ear, you will probably find a thickening of cartilage. The facts of embryological development show that this thickening is comparable with the top end of a pointed ear that is folded over.

Cut off the left **pinna** close to the skull bones of the cat. The **external acoustic (auditory) meatus** is the hole or opening in the side of the head above the **tympanic (auditory) bulla** (see Fig. 1-4). Observe the cartilage that lines the bony cavity. The **skin ectoderm** covering the pinna continues and lines the cavity and near its inner end gives rise to the **ceruminous, or wax, glands,** which are too small to be seen in ordinary dissection. Cut this cartilage and bone away until the meatus enlarges and

the **eardrum,** or **tympanic membrane,** can be seen. It arises by the skin ectoderm, infolding to form the lining of the meatus with the endoderm lining from the pharynx pushed up through the **auditory (eustachian) tube,** which lines the middle ear vestibule and forms the **inner layer** of the tympanic membrane.

2. The **middle ear** consists of the vestibule, or cavity, which contains the malleus, incus, and stapes and also the inner half of the tympanum. The vestibule and eustachian tube represent the inner portion of the first embryonic pharyngeal pouch or gill cleft. Find the **opening of the auditory tube** (Fig. 6-3) in the lateral wall of the nasopharynx partly surrounded by the hamular process of the sphenoid bone, which can be felt by the tip end of the finger. The edge of the opening is often thickened and colorless. Insert a fine wire or probe into the opening and push it up into the auditory tube. Now cut away the bone covering the wire or probe, holding the scalpel in a horizontal position. The auditory tube and vestibule permit the air pressure to be equalized on the two sides of the eardrum, which then vibrates more easily. If you hold your own nose and swallow, you increase the air pressure from the pharynx up through the auditory tube against the eardrum; it does not vibrate so easily, and you do not hear so well.

The bones of the middle ear, the **malleus,**

incus, and **stapes,** are important in the process of hearing, since they transmit the stimulus produced by the sound waves. These ear bones originate from the **visceral, or branchial, arches** of the embryo: the malleus and incus from the **mandibular,** or **first arch,** and the stapes from the second, or **hyoid arch.** They are difficult to expose in dissecting because they are surrounded by hard bone, are small, and are easily damaged. Cut away the bone at the **tympanum** and loosen the rostral half of the tympanum. The **malleus** can be seen with one end on the inner side of the eardrum, with the large end caudal to the membrane, where it articulates with the **incus.** The incus in turn articulates with the **stapes,** which contacts the **oval window.** Now cut away the caudal wall of the tympanum and locate the incus and the stapes. Fine folds of the lining of the vestibule support these bones; therefore, it is somewhat difficult to identify them. If you succeed in removing these three bones intact, you may consider yourself skillful and perhaps lucky also. They are easier to dissect intact on a dry skull.

Now you can pass a fine wire up through the auditory tube, the vestibule, past the tympanum, and out the remainder of the external acoustic meatus.

Cut away the ventral wall of the **tympanic bulla,** and expose its **cavity.** The outline of the cavity varies with the depth at which it is cut. The oval thickened area of the dorsolateral wall of the bulla contains the parts of the **inner ear** within the petrous part of the temporal bone. A small opening in the dorsolateral wall of the bulla connects with the vestibule. It is believed that the tympanic bulla amplifies sound much the same as the body of a violin, cello, or bass viol. The bulla probably accounts in part for the greater keenness in hearing in the cat than that in man, who has no bulla.

3. The **inner ear** is composed of the **semicircular canals,** the **cochlea,** the **sacculus,** and the **utriculus.** All of these parts are embedded in the **petrous part of the temporal bone,** which is the hardest bone in the body. In fact it is so hard that it is quite difficult to break. It is endochondral, or preformed in cartilage, in its development. Cut or break away parts of the petrous temporal bone as best you can, and you may be lucky enough to see parts of the cavities of the **semicircular canals** and **cochlea.** A hand lens is usually necessary in identifying

these parts. In each ear there are three semicircular canals, which lie in different planes, and when the animal changes position, the endolymph shifts and stimulates the nerve endings differently in the ampullae. Because of this stimulus on the nerve endings in the ampullae, the animal is able to keep its balance, or **equilibrium.** The **cochlea** is the spiral, or snail-shell–shaped, organ. The highly specialized cells within constitute the **organ of Corti.** These cells are specialized in such a manner that they respond to specific stimuli produced by a definite wavelength sound vibration. It is not definitely known how the ear functions. The sacculus and utriculus are small cavities with which the semicircular canals and the cochlea connect. All cavities of the inner ear, or labyrinth, are lined by the infolding of the skin ectoderm. For a time in the embryo, these cavities are connected with the skin by a tube known as the endolymphatic duct, as in the adult *Squalus acanthias* shark. However, this tube constricts off in the cat and man, and the various parts differentiate.

STRUCTURAL DEVELOPMENT OF EAR

The structures of the **inner ear** are the first to begin formation in the embryos of cat and man. Then the **middle ear** forms, and lastly the **external ear** forms. This is the same order in which these parts form phylogenetically, or in the development of the group of vertebrates. In the **lowest fishes,** as in the dogfish shark *(Squalus acanthias),* **only an inner ear** forms. In the amphibians, as in the frog, there is **only an inner** and a **middle ear** and no external ear. The middle ear bones are represented only by the columella and are not separate malleus, incus, and stapes bones. When structures of a higher-classed animal, such as the cat and man, pass through stages in their embryological development similar to those in the adults of lower members of the same group, it is designated as **recapitulation.** Recapitulation is interpreted as supporting the theory of evolution. Since the semicircular canals form first embryologically and phylogenetically, it is believed that they existed a long time before hearing was established.

In the cat there are rostral, dorsal, caudal, and ventral **auricular muscles** on the side of the skull by which the ear is moved. This enables the cat to catch the sound more easily and efficiently, which is important in avoiding enemies or in detecting other small animals to catch and use as food. These muscles are **vestigial** in man, but in nearly every schoolroom there is a young boy who can move or wiggle his ears for the entertainment of others.

There are many rudimentary, degenerative, or vestigial organs in man that have little or no known function. Dr. Leslie B. Arey in the seventh edition of *Developmental Anatomy*, published by W. B. Saunders Co. in 1965, states that "**over 100** such organs have been listed for man, among such are the **coccyx, appendix, body hair, wisdom tooth,** and **ear muscles.**" It is believed by most anatomists that an organ that is not used tends to degenerate and atrophy. This is the usual explanation of vestigial organs.

Name _____

Date _____

REVIEW QUESTIONS ON EAR

1. State the two principal functions of the ear.

2. What is the difference between an aortic arch and a branchial arch?

3. What is the explanation of the area of thickened cartilage at the upper edge of the human ear?

4. Name the principal structures of each of the three main parts of the ear. (Fig. 6-3)

5. What is believed to be the function of the tympanic, or auditory, bulla?

6. Why does swallowing often interfere with the acuteness of man's hearing?

7. State the embryological origin of each of the bones of the middle ear. (Fig. 6-3)

8. What is the auditory (eustachian) tube called in the early embryo?

9. Does air come up from the pharynx, eustachian tube, and vestibule of the middle ear, to the tympanic membrane? If so, what is the result?

10. Name the bones of the middle ear. Why are they sometimes called the hammer, anvil, and stirrup?

11. Because of the complexity of the cavities of the inner ear, what is the general name of all the cavities?

12. What is the function of the semicircular canals?

13. What primitive germ layer lines each of the cavities of the inner ear?

14. What is the nature of the petrous temporal bone?

15. In what part of the ear is the most essential organ of hearing found?

16. Within what part of the temporal bone is the organ of Corti?

17. Name five senses of cat or man.

18. Hold your nose and swallow. Air is forced from the pharynx up the eustachian tube against the tympanum, or eardrum, and you can feel the sensation. Can you hear well during the process? Why?

19. In what part of the ear are the physical stimuli transferred or transformed into nervous stimuli? In other words, where are the receptive ends of the auditory nerve located?

20. Trace the stimulus produced by sound waves entering the acoustic meatus until it reaches the brain.

SEVEN
Brain of the cat

BRAIN, MEDIAN SAGITTAL (Fig. 7-1)

Carefully remove the left half of the brain from the skull, previously examined. Embryology shows that the brain is really the enlarged end of the spinal cord. There are five principal parts, or lobes, that constitute the brain, and there are several parts in each lobe. Each of these lobes will be considered, but only the parts of each that may be seen in the median sagittal section will be mentioned.

1. The **cerebrum (telencephalon)** is the most rostral of the lobes and occupies the largest portion of the cranial cavity. It is covered by the frontal and parietal bones and rests on the diencephalon and mesencephalon. The following structures are some of the parts of the cerebrum that may be seen.

a. The convolutions consist of **gyri** (ridges) and **sulci** (grooves).

b. The **olfactory bulbs (rhinencephalon)** lie against the cribriform plate, and some of the small olfactory nerve fibers may be seen entering the foramina if the lobe is gently pushed back.

c. The **corpus callosum** may be identified as follows. Each hemisphere of the cerebrum contains a cavity, the paracele. The corpus callosum forms the dense white roof of this cavity near the median plane. This is composed of many nerve fibers extending from one hemisphere to the other. Below the paracele is the **fornix,** which is also composed of transverse nerve fibers closely packed together.

d. The **septum pellucidum** is a thin, double partition that lies in a vertical position in the median plane and separates the two paraceles (cavities) of two hemispheres of the cerebrum.

2. The **lower thalamic region (diencephalon)** is the second main lobe of the brain and is composed of the following parts.

a. The **intermediate commissure** appears as a solid, circular area extending transversely near the center of the cavity of the **diacele,** or **third ventricle.** The intermediate commissure also is composed of many nerve fibers.

b. The **epiphysis (pineal gland,** pineal body, or vestigial third eye) is small and dorsal to the intermediate commissure. It constitutes a part of the dorsal wall, or roof of the third ventricle. In the early embryological development the **epiphysis** is actually on the dorsal surface of the brain, but the **cerebrum** enlarges and folds back, covering the epiphysis, which is still really on the dorsal surface of the brain.

c. The **infundibulum** is a ventral projection from the floor of the diencephalon. It is the caudal portion of the pituitary, or hypophysis, which lies in the sella turcica of the sphenoid bone. The rostral portion of the pituitary arises from the roof of the stomodeum, or mouth (**Rathke's pouch**).

d. The **optic chiasma** is rostral to the pituitary and is composed of the optic nerve fibers, most of which cross one another in this area.

e. The **tuber cinereum** is a small rounded area between the optic chiasma and the base of the pituitary gland to which the infundibulum is attached.

f. The **interventricular foramen** is the opening that connects the cavity of the diencephalon (the **diacele,** or **third ventricle**) with the lateral ventricle (paracele) of each hemisphere of the cerebrum. The **choroid plexus** is an infolded part of the roof of the diacele extending through the **interventricular foramen** into each paracele. It is dark because of the blood it contains. The cerebrospinal fluid is produced at the choroid plexuses of each ventricle. It flows through foramina in the roof of the fourth ventricle to the **subarachnoid space.**

143

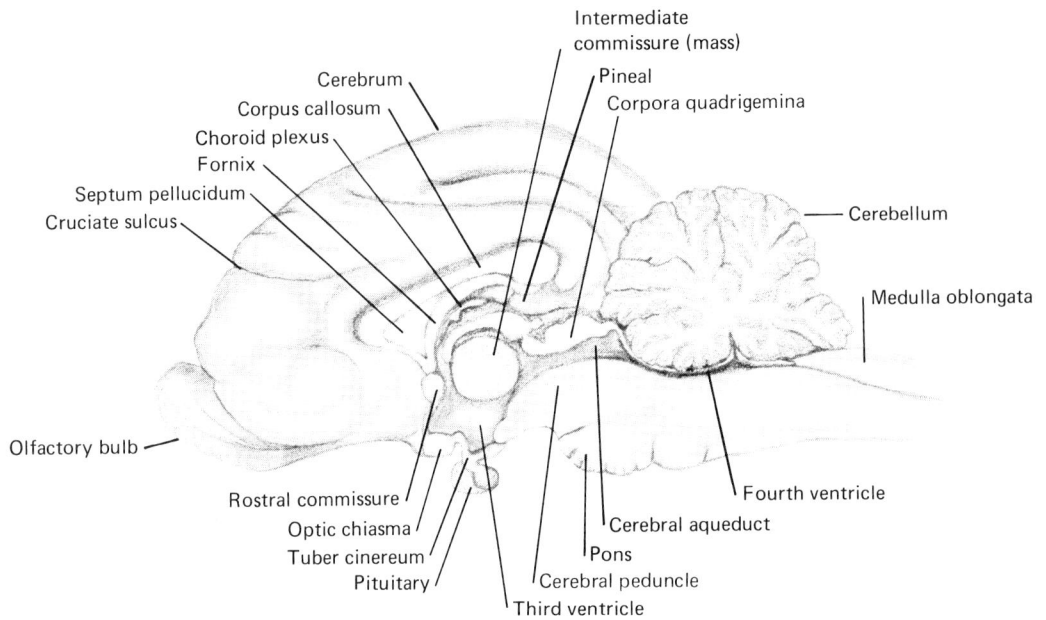

Fig. 7-1. Brain, median sagittal view.

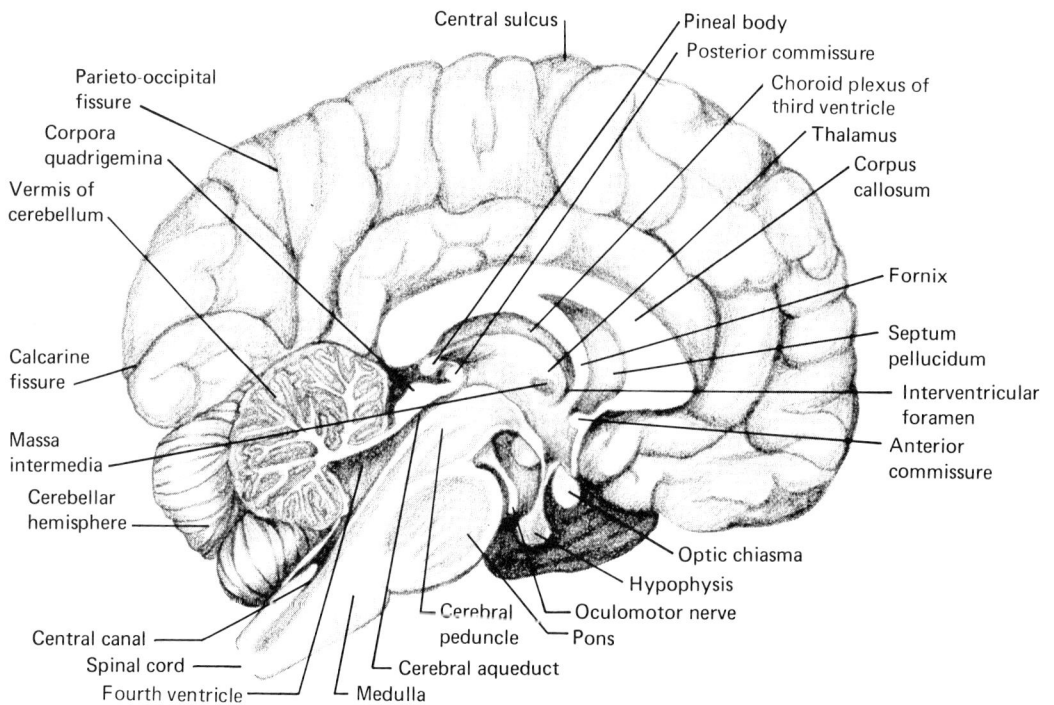

Fig. 7-2. Human brain, median sagittal view.

144

3. The **midbrain (mesencephalon)** has two lobes (colliculi) on each side, one behind the other, caudal to the pineal gland. Since there are four of these lobes, they are also given the name **corpora quadrigemina.** The rostral pair constitutes a relay center for sight and the caudal pair, a relay center for hearing. The cavity below these lobes is very small and is known as the **cerebral aqueduct.**

4. The **cerebellum (metencephalon),** as it lies in the skull, is partially separated from the cerebrum by the tentorium, which is really an ossified portion of the dura mater. When cut in median sagittal section, the cerebellum shows a characteristic branching structure, and since it somewhat resembles a tree, it is called the arbor vitae. The cavity ventral to the cerebellum is the **fourth ventricle,** and it also extends caudally into the medulla oblongata. Lying transversely ventral to the rostral end of the medulla is the **pons.**

5. Most of the roof of the **medulla oblongata (myelencephalon)** is very thin and often is destroyed in dissecting before being identified. A part of the roof folds into the fourth ventricle, forming a **choroid plexus** on each side, which extends to the inner surface of the cerebellum. The lateral walls are thicker, and the floor is thickest of all. The caudal limit of the medulla is at the foramen magnum (see Fig. 1-4) of the skull, where it joins the spinal cord. So far as position and structure are concerned, the medulla is a transition between the brain proper and the spinal cord.

Compare Figs. 7-1 and 7-2. Note five similarities and five differences between the brains of the cat and man.

BRAIN, DORSAL (Fig. 7-3)

If you worked in pairs on the pharynx and the medial view of the brain, as previously stated, you now have an intact skull from which the entire brain may be removed. It is better not to cut the head from the body until the dorsal skull bones have been removed.

A good pair of bone shears is needed for removing the bones of the roof of the skull in order to remove the brain intact. Begin at the rostral limits of the frontal bones and cut through into the frontal sinuses. Continue clipping small pieces of bone away without injuring the brain until the roof is entirely removed. If the brain is too soft, the entire skull with the brain should be hardened in formalin unless disintegration has gone too far.

Remove the hardened brain from the skull by first lifting up the medulla, then the cerebellum; lift forward until the entire brain is loosened and removed. Special care must be exercised so as not to injure the pituitary gland. Re-examine Fig. 7-1 to determine how to cut down with a scalpel behind and under the pituitary so as to remove the brain uninjured.

The **meninges** are the covering membranes of the brain. They will be considered, beginning with the outer one.

1. The **dura mater,** which was previously mentioned in the discussion of the cerebellum, mostly lies against the bones. In the cat, part of it is ossified to form the tentorium; but this is not so in man. It incorporates the periosteum.

2. The **arachnoid,** the middle layer, contains the blood vessels, which are in the subarachnoid space.

3. The **pia mater,** or inner layer of the meninges, may be seen as the thin membrane that dips down into the sulci, or grooves, of the convolutions of the brain and carries blood vessels into the brain substance.

Next, examine the dorsal view of the cerebrum. The **longitudinal,** or **median, fissure** separates the two hemispheres; each hemisphere has three longitudinal folds, or gyri, on its dorsal surface, which are separated from one another by longitudinal grooves, or sulci. The three principal gyri on each half of the brain are named, beginning near the median line, as the (a) **marginal,** (b) **ectomarginal,** and (c) **ectosylvian.** The four principal sulci on each half are named from the median fissure laterally as the (a) **lateral,** which is in two parts, the caudal portion turning laterally near the cerebellum, (b) **suprasylvian,** which turns laterally near each end, (c) **cruciate sulcus,** which extends transversely on each side near the olfactory bulb and begins at the median fissure, and (d) **sylvian fissure,** which is at the greatest width of the cerebrum as it comes to the dorsal surface.

The **diencephalon** and the **mesencephalon** are comparatively small in the cat and have been covered by the enlargement and growth of the caudal portion of the cerebrum, and hence they are not seen in the dorsal view.

The **metencephalon (cerebellum)** is characterized by having many small irregular folds,

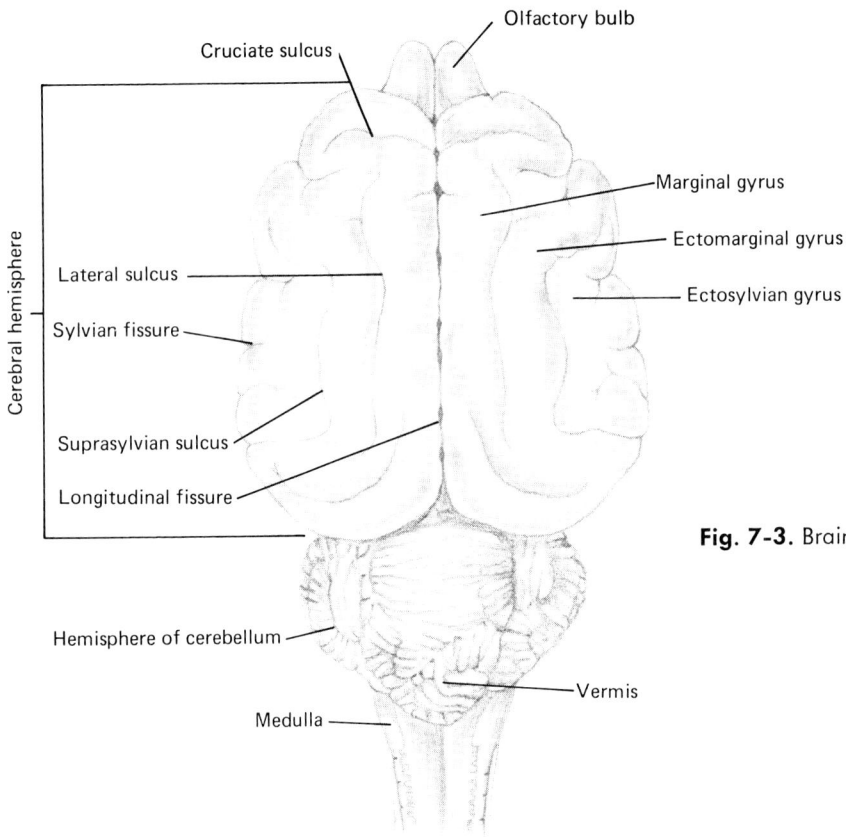

Fig. 7-3. Brain, dorsal view.

Olfactory bulb

Cruciate sulcus

Marginal gyrus

Ectomarginal gyrus

Ectosylvian gyrus

Cerebral hemisphere

Lateral sulcus

Sylvian fissure

Suprasylvian sulcus

Longitudinal fissure

Hemisphere of cerebellum

Vermis

Medulla

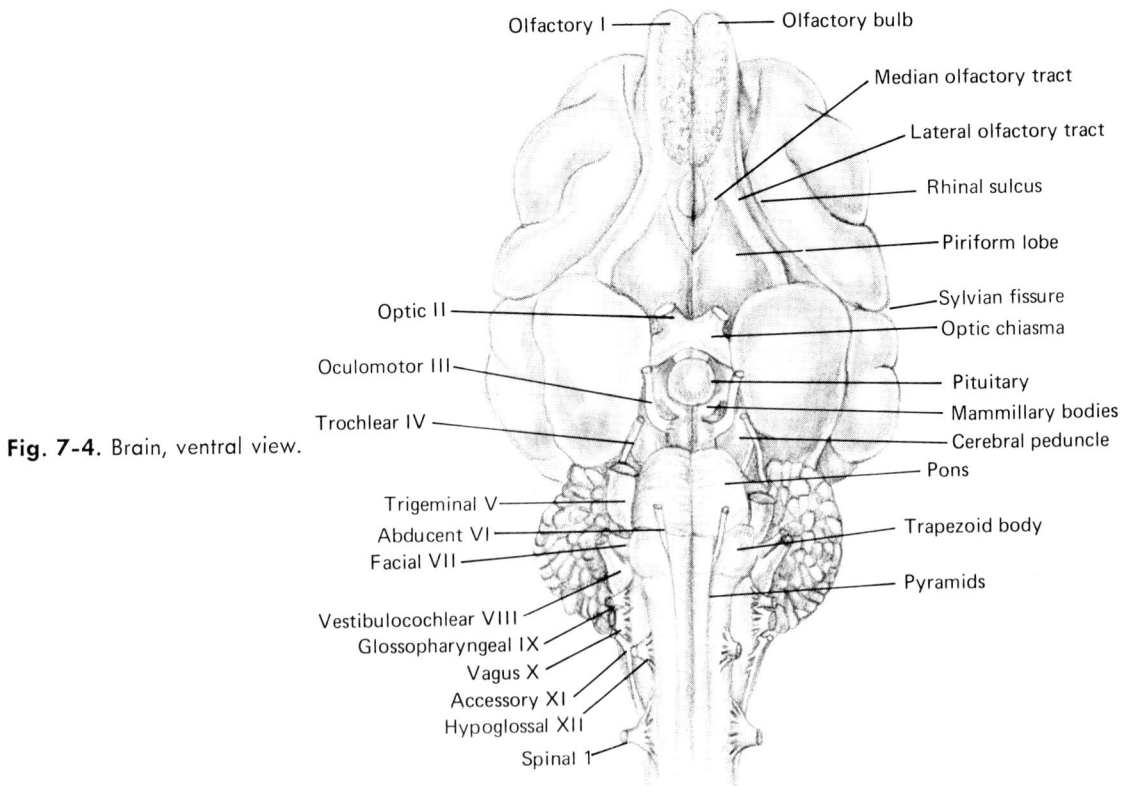

Fig. 7-4. Brain, ventral view.

Olfactory I

Olfactory bulb

Median olfactory tract

Lateral olfactory tract

Rhinal sulcus

Piriform lobe

Optic II

Sylvian fissure

Optic chiasma

Oculomotor III

Pituitary

Trochlear IV

Mammillary bodies

Cerebral peduncle

Pons

Trigeminal V

Abducent VI

Trapezoid body

Facial VII

Vestibulocochlear VIII

Pyramids

Glossopharyngeal IX

Vagus X

Accessory XI

Hypoglossal XII

Spinal 1

146

or gyri. These are arranged in three principal groups. The median portion is the **vermis**, which has a **lateral lobe**, or **cerebellar hemisphere**, on each side.

The **myelencephalon** (**medulla oblongata**) may be seen from the dorsal view as the rapidly narrowing structure projecting from under the caudal fold of the cerebellum and is continuous with the spinal cord.

Color each of the lobes shown.

BRAIN, VENTRAL (Fig. 7-4)

The ventral view of the cat brain appears more complex than the dorsal. This is largely because the brainstem is exposed better with the bases of the cranial nerves attached to it.

Each **olfactory bulb** and its **tract** may now be observed with its **lateral roots**. Cranial nerve **I** (**olfactory**) fibers extend forward from each olfactory bulb through the cribriform plate and end on the turbinate bones. The **postrhinal fissure** may be traced along the lateral side of each **olfactory bulb** to the lateral surface, where it is usually continuous with the **sylvian fissure**. The crossing of the **cranial nerve II** (**optic**) fibers forms the **optic chiasma**. The **pituitary body**, or **hypophysis**, is the knoblike projection immediately caudal to the chiasma; dorsal to it is the **tuber cinereum**. The **mammillary bodies** lie caudal to the tuber cinereum.

Cranial nerve III (**oculomotor**) arises near the midventral line caudal to the mammillary bodies. Lateral to the tuber cinereum, the mammillary bodies, and the third cranial nerve, there is a large oval swelling, the **piriform lobe** of the cerebrum. On each side of the mammillary bodies and the piriform lobe is a depressed area that is a part of the **cerebral peduncle**. Near its lateral limit, **cranial nerve IV** (**trochlear**) may be seen. Farther caudal is a transverse enlargement, the **pons,** in which are tracts, or association fibers, of the cerebellum. Two longitudinal bodies flanking the midline ventral in the medulla and caudal to the pons are the **pyramids**. Lateral to the pyramid on each side and caudal to the pons is **cranial nerve VI** (**abducent**). From the lateral portion of the pons arises **cranial nerve V** (**trigeminal**). The **trapezoid body** lies caudal to the pons. Laterally from the trapezoid area are **cranial nerves VII** (**facial**) and **VIII** (**vestibulocochlear**).

Cranial nerves IX (**glossopharyngeal**) and **X** (**vagus**) arise near the lateral surface of the medulla slightly caudal to the eighth nerve. **Nerve XI** (**spinal accessory**) arises from several fibers along the lateral edge of the medulla and from the spinal cord. It is caudal to nerve X. **Cranial nerve XII** (**hypoglossal**) arises on the ventral surface of the medulla and appears at its base very much the same as the first cervical nerve.

The floor of the embryonic brain (brainstem) with the bases of the cranial nerves changes least in development. The dorsal and lateral walls thicken most, particularly those of the first and fourth lobes. The principal bending of the brain is correlated with the erect position of man.

ARTERIAL SUPPLY TO BRAIN (Fig. 7-5)

Several arteries may be seen on the ventral surface of the well-injected brain. Identify these arteries on your specimen and on Fig. 7-5.

The **vertebral artery** extends forward on each side along the neck, passing through a series of transverse foramina in the transverse pro-

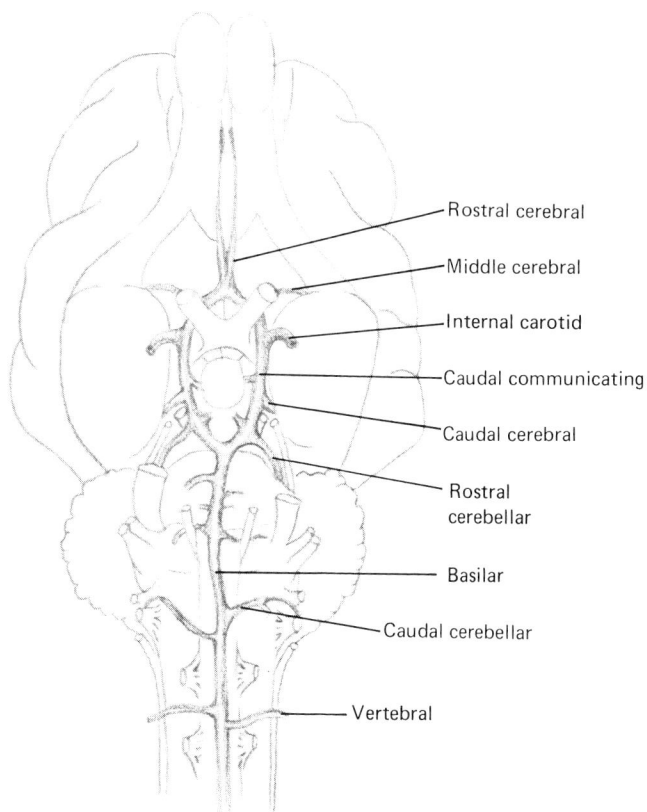

Rostral cerebral

Middle cerebral

Internal carotid

Caudal communicating

Caudal cerebral

Rostral cerebellar

Basilar

Caudal cerebellar

Vertebral

Fig. 7-5. Arteries of the brain.

cesses of the cervical vertebrae. The arteries converge, passing through the foramen magnum, and unite into the basilar artery. They may be joined by branches of the occipital artery.

The **basilar artery** extends forward against the median ventral surface of the medulla and the pons. Upon reaching the mammillary bodies it divides and passes on each side of the bases of the hypophysis, or pituitary, and the optic chiasma, forming the **arterial circle.**

The **pituitary** is considered the **master endocrine gland,** since it controls the secretions of several hormone-producing glands. The thyroid, parathyroid, pancreas, suprarenal, ovary, and testis are other endocrine glands.

Along the course of the basilar artery are the **caudal cerebellar arteries,** and, more rostrally, is a pair of **rostral cerebellar arteries.**

At the rostral end of the basilar artery a pair of larger arteries, the **caudal cerebral arteries,** are given off. The arterial circle receives the **internal carotid artery** on each side and passes rostrally to the chiasma, where it forms a complete circle and gives off a pair of **middle cerebral arteries** to supply much of the ventral surface of the cerebrum. **Smaller arteries** branch from near the median line to extend to the bases of the **olfactory lobes.** On each side of the pituitary, a **communicating artery** unites the caudal cerebral and the internal carotid arteries.

The **vertebral** and **internal carotid arteries** are the main supply of blood to the brain. The **dura mater** is supplied by **rostral, middle, and caudal meningeal arteries.** The veins of the brain drain into the sinuses of the dura mater and then by emissary veins to the external veins.

SOME DIFFERENCES IN BRAINS OF THE CAT AND MAN

1. The cerebrum of man is relatively larger and has many more complex convolutions than that of the cat.

2. The brain of man is bent more than that of the cat.

3. The olfactory bulbs in the cat are relatively larger and more protruding than those in man.

4. The walls of the cerebrum and cerebellum in man are relatively thicker than they are in the cat, and they contain more cells and synapses.

5. The hemispheres of the cerebellum are more definitely differentiated from the central vermis in man than they are in the cat.

6. In man the enlarged cerebral lobes of the brain grow over and cover the olfactory bulbs and the cerebellum. This does not occur in the cat.

7. The median commissure in the third ventricle of the cat is relatively larger than that in man.

REVIEW QUESTIONS ON BRAIN

1. Give the common and scientific names of each of the lobes of the cat brain.

2. Name the bones of the skull that help form the cranial cavity. (Fig. 1-6)

3. Name the three meninges.

4. What constitutes the brainstem?

5. What is the arterial circle?

6. Name the cavities of the brain and state the location of each.

7. List the 12 pairs of cranial nerves and state the location where each arises from the brainstem.

8. Trace the stimulus of a sound wave that strikes the pinna until it reaches the brain.

9. What are the principal differences in the cat and human brain?

10. What are sulci? Name three.

11. What are gyri?

12. What two arteries unite to form the basilar artery?

13. Where is the pituitary gland located? What is its function? Name the other endocrine glands.

14. Explain and locate two choroid plexuses.

15. What are the lateral ventricles (paraceles)?

16. What is the interventricular foramen, and what is its purpose?

17. What arteries give the main supply of blood to the brain?

EIGHT
Spinal cord and peripheral nerves of the cat

Typically, there is one pair of nerves for each vertebra of the spinal column. Although there are only seven cervical vertebrae, there are eight pairs of cervical nerves, one pair being immediately caudal to the skull. The caudal (coccygeal) nerves are greatly reduced in number.

Each pair of nerves has dorsal and ventral roots, branches on each side from the spinal cord. Each nerve thus supplies a **myomere** (**myotome**, or **muscle plate**); however, the nerves that supply the limbs become larger and more complicated by branching and reuniting with one another. In this manner **plexuses** are formed not only in the limbs but also in many of the internal organs. The nerve fibers forming the plexuses are spoken of as **anastomosing** with one another. Blood vessels also **anastomose.** Moreover, by counting the number of basic nerves that supply a limb, one has some idea of the number of **mesodermal somites,** or muscle plates, that go into its embryological development. The various muscles supplied by the branches of the same nerve are believed to have originated from the same mesodermal somite.

NERVES OF BRACHIAL PLEXUS (Fig. 8-1)

Bisect and reflect the pectoralis muscles on both sides of the thorax and muscles of the neck. **Cervical nerves 1 to 5** branch from the spinal cord and supply only the neck. The ventral branches of **cervical nerves 6, 7,** and **8** together with the **first thoracic** come from the spinal cord and anastomose to form the **brachial plexus.** From this plexus nerves supply the shoulder, forelimb, and thoracic wall. Leave

the nerves and blood vessels in position but pull away the fat and connective tissue surrounding them so that they may be identified.

Much time and patience are required to make a satisfactory dissection of the brachial plexus. Take up each nerve as described from its base to the structures it supplies to be sure that it is accurately identified. You should be able to demonstrate any particular nerve to the in-

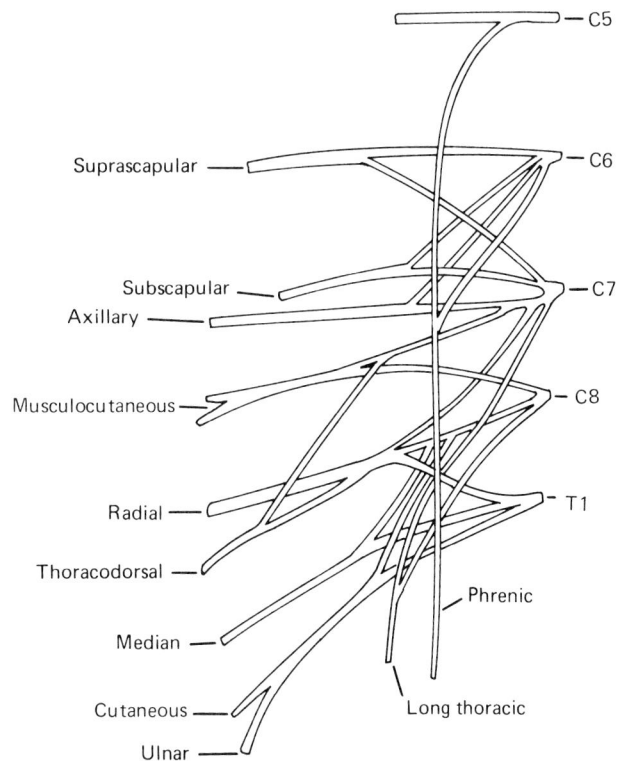

Fig. 8-1. Nerves of brachial plexus.

structor. If the work is done thoughtfully, some variations from the descriptions will be observed.

The following nerves are to be identified (Fig. 8-1).

1. The **suprascapular nerve** is the most cranial of the brachial plexus group; it comes from cervical nerves 6 and 7 and extends over the cranial edge of the scapula to supply the supraspinatus and infraspinatus muscles. Locate the upper edge of the scapula with a dissecting needle. The nerve is closely associated with the arteries of the thyrocervical axis.

2. The **subscapular nerve** arises from cervical nerves 6 and 7 caudal to and parallel with the suprascapular nerve. It can be identified as it enters the subscapular muscle. It may be double.

3. The **thoracodorsal nerve** arises from cervical nerves 7 and 8. It has few branches and supplies the latissimus dorsi muscle.

4. The **phrenic nerve** is often included with the brachial plexus. It originates from the bases of cervical nerves 5 and 6 as two slender roots that unite and pass into the thoracic cavity ventral to the base of the lungs to supply the diaphragm. Find it at the base of the lung and trace it both ways.

The question might arise concerning why the phrenic nerve should originate at the place where it does and pass through the thoracic cavity to get to the diaphragm. Why should it not originate from nerves of the lower thoracic region close to the diaphragm? The explanation is that the diaphragm (septum transversum) in the young embryo is in the region of the bases of the sixth and seventh cervical nerves, and when these nerves grow out, they become attached to the diaphragm. Then later, when the diaphragm migrates down past the thoracic region, it takes the phrenic nerves along with it. This explanation also applies to man.

5. The **axillary nerve** arises from cervical nerves 6 and 7 and courses laterally between the teres major and subscapularis muscles. It continues behind the head of the humerus and supplies the teres major, teres minor, and deltoideus muscles.

6. The **musculocutaneous nerve** arises from cervical nerves 6 and 7, continues to the coracobrachialis muscle, passes down the medial surface of the biceps, and continues on to the brachialis and to the skin of the forearm.

7. The **radial nerve** is the largest of all the nerves in this region. It lies along the axillary artery and has roots from cervical nerves 7 and 8 and thoracic nerve 1. It supplies the medial surface of the large head of the triceps. It passes along the musculospiral groove of the humerus, where it divides into superficial and muscular branches. It supplies the extensors of the elbow, carpus, and digits.

8. The **median nerve** has long roots that unite at the same level as the head of the humerus. It passes through the supracondylar foramen and supplies the flexors of the carpus and digits.

9. The **ulnar nerve** is the largest of the remaining nerves. It usually lies close to the axillary and brachial arteries. It is the most caudal of the nerves that extend down the brachium to the elbow. It passes caudal to the supracondylar foramen and across the olecranon process to the flexor carpi ulnaris muscle and other muscles of the forearm. It also supplies the flexors of the carpus and digits.

10. The **medial cutaneous nerve** arises from the ulnar nerve and continues close to the base of the ulnar nerve to the distal half of the brachium. Here it passes to the skin on the ulnar side of the forearm.

11. The **external thoracic nerves** arise from the eighth cervical and pass ventrally close to the ventral thoracic artery and vein to supply the underside of the pectoralis major muscle.

12. The **long thoracic nerve** arises from cervical nerves 7 and 8. It passes near the long thoracic artery to the underside of the pectoralis minor and lateral surface of the serratus ventralis muscles. These nerves are small and extend through the loose connective tissue, and thus some difficulty may be encountered in their identification.

AUTONOMIC NERVOUS SYSTEM

The **thoracolumbar (sympathetic)** portion of the **autonomic nervous system** (see Fig. 8-3) is dissected here because of its close association with the structures just dissected, but it will be considered again later. It consists of the ganglionated nerve chains, or trunks, their branches, plexuses, and many small ganglia. Dissect away the thin peritoneum from the left thoracic wall, about a centimeter lateral to the spinal column. The nerve chain may be seen as a fine white line behind the peritoneum. The ganglia are small enlargements on the nerve chain, and from them small fibers join the spinal nerves of the body wall. Fibers are received from the thoracolumbar spinal cord in the region of each segment.

Dissect out the **left nerve trunk** forward to the **first thoracic ganglion**, which is immediately dorsal to the **subclavian artery** (see Fig. 5-1). A nerve fiber passes forward on each side of this artery to the **caudal cervical ganglion**, which is smaller and more medial in position. Because of its shape, the first thoracic ganglion is called the **stellate ganglion** (also cervicothoracic). Continue to dissect carefully the nerve trunk, or chain, forward as it passes dorsal to the **brachiocephalic vein** (see Fig. 4-1) and turns more medial to come into close association with the **vagus nerve**, forming the **vasosympathetic trunk**. Both are in the same connective tissue sheath. This trunk continues to the upper region of the neck to what appears to be one large ganglion. Separate the sympathetic trunk from the vagus nerve. The ganglion of the **sympathetic trunk** is the **cranial cervical ganglion.**

Dissect out the sympathetic nerve trunk in the region of the diaphragm and find the **greater splanchnic nerve fibers** going to the **large celiac ganglion** and the **cranial mesenteric ganglia** in the dorsal mesentery close to the bases of the **celiac** and **superior mesenteric arteries**. From these ganglia many fine nerve fibers branch to most of the abdominal viscera, and together they constitute the solar plexus (celiac and cranial mesenteric plexuses). Dissect out the nerve chain posterior to the kidneys and find the branching nerve fibers going to the **caudal mesenteric ganglion** close to the **caudal mesenteric artery**. Both sides are represented in a single ganglion. From each side a hypogastric nerve goes to the pelvic viscera. The sympathetic nervous system transmits stimuli that accelerate the heart, constrict blood vessels of the skin, inhibit the gastrointestinal activity, and dilate the bronchi.

The **vagus nerve** is a part of the **parasympathetic** portion of the autonomic nervous system. The vagus is a cranial nerve arising from the medulla oblongata and supplying the heart, lungs, stomach, and small intestines. In the early embryo the aforementioned organs are ventral to the medulla, and the vagus branches become attached to these organs; as they migrate down into the thoracic and abdominal cavities, they take these nerve endings along with them. This explanation is much the same as that for the phrenic nerve and the diaphragm, and it applies equally for the cat and man.

NERVES OF LUMBOSACRAL PLEXUS (Fig. 8-2)

Since the pubic bones have been separated at the symphysis, cut off the urinary bladder, urethra, and rectum from their attachments in the lumbar and sacral regions and reflect them. Remove the arteries and veins from this region and carefully dissect away the muscles and connective tissue until the bases of the nerves are exposed as they come from the vertebrae.

Identify the centra of the fourth to the seventh lumbar vertebrae. Lateral to these are many lumbar and sacral nerves. Dissect the psoas

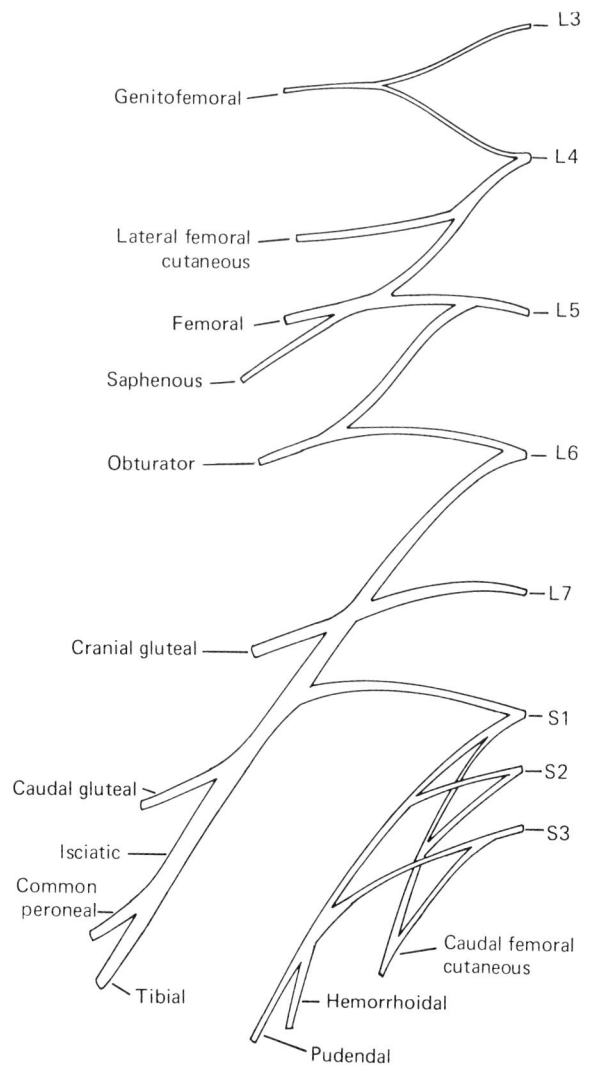

Fig. 8-2. Nerves of lumbosacral plexus.

muscles as you look for the nerves. The principal nerves of the lumbosacral plexus are as follows.

1. The **genitofemoral nerve** is small and arises from lumbar nerves 3 and 4. It passes caudolaterally medial to the psoas muscles along the external iliac vessels to the inguinal region.

2. The **lateral femoral cutaneous nerve** is from lumbar nerve 4. It courses laterally dorsal to the psoas muscles and supplies the cranial and lateral surfaces of the thigh.

3. The **femoral nerve** is large and enters the cranial part of the thigh by passing through the iliopsoas. It is formed from lumbar nerves 4 and 5. As it enters the thigh it gives off the saphenous nerve, which accompanies the saphenous vessels along the medial and caudal surfaces of the thigh.

4. The **obturator nerve** arises from lumbar nerves 5 and 6. It is parallel to the femoral nerve at first but turns caudally along the shaft of the ilium to pass through the obturator foramen. It supplies the medial muscles of the thigh.

5. The **ischiatic (sciatic) nerve** was observed when dissecting the muscles of the hip and thigh, but its origin for the right pelvic limb is now to be dissected from the ventral surface. Dissect off the muscles from the ventral surface of the lumbosacral junction. The sciatic nerve is very large and arises from lumbar nerves 6 and 7 and sacral nerve 1.

6. The **cranial gluteal nerve** arises with the sciatic nerve. It passes over the greater sciatic notch to the muscles of the hip.

7. The **caudal gluteal nerve** arises with the sacral nerve and, after traversing the greater sciatic notch, supplies the muscles of the caudal gluteal area.

Find the sciatic and the place where it gives off a fairly large muscular branch to the inner surfaces of the upper portions of the biceps femoris, semitendinosus, and semimembranosus muscles. The sciatic passes down medial to the biceps femoris and gives off more muscular branches and two crural cutaneous nerves. The sciatic nerve branches in the popliteal space into the more lateral **common peroneal nerve,** which passes along the medial surface of the lower end of the biceps and the lateral surface of the lateral head of the gastrocnemius, and the **tibial nerve,** which passes between the medial and lateral heads of the gastrocnemius.

8. The **caudal femoral cutaneous nerve** originates from sacral nerves 1, 2, and 3. It is small and has few branches. It comes to the lateral surface between the upper ends of the caudofemoralis and the biceps femoris. It crosses lateral to the biceps and supplies the skin of the caudal surface of the thigh. It may be associated with the caudal gluteal nerve.

9. The **pudendal nerve** arises from the sacral nerves, which unite lateral to the first caudal vertebra. It then turns medially to supply the external genitalia and tissues lateral to the rectum and the area of the ischial arch.

10. The **hemorrhoidal nerve** is a branch of the pudendal nerve and supplies the anal area. There are other small fibers of the sacral plexus that are not considered to be of significance for this manual.

TWO MAIN DIVISIONS OF AUTONOMIC NERVOUS SYSTEM (Fig. 8-3)

"Autonomic" means acting independently of volition. The autonomic nervous system regulates the involuntary reflexes especially concerned with nutritive, vascular, and reproductive activities. We can get only a general idea of this system in gross dissection since most of the nerve fibers and ganglia are small and require a special technique to identify. There are two main divisions: (1) **sympathetic (thoracolumbar)** and (2) **parasympathetic (craniosacral).**

1. The **thoracolumbar division** (previously mentioned) consists of nerve fibers that arise from the spinal cord from the thoracic and lumbar regions and pass laterally on each side to the sympathetic nerve cord or trunk, about a centimeter from the spinal column. From here one group of fibers goes to the cutaneous blood vessels and sweat glands by way of the spinal nerves, and the other group goes to the heart, lungs, and the various organs of the digestive and urogenital systems. In the abdominal cavity many of these fibers pass through the celiac, cranial, or caudal mesenteric ganglia.

2. The **craniosacral nerve fibers** arise from one of two groups, either from the brain or from the spinal cord in the sacral region. Those fibers from the brain are parts of cranial nerves III, VII, IX, and X, and in the sacral region they are from sacral nerves 2, 3, and 4. These nerves are characterized by having ganglia on or near each organ supplied.

To a certain degree the **thoracolumbar** and

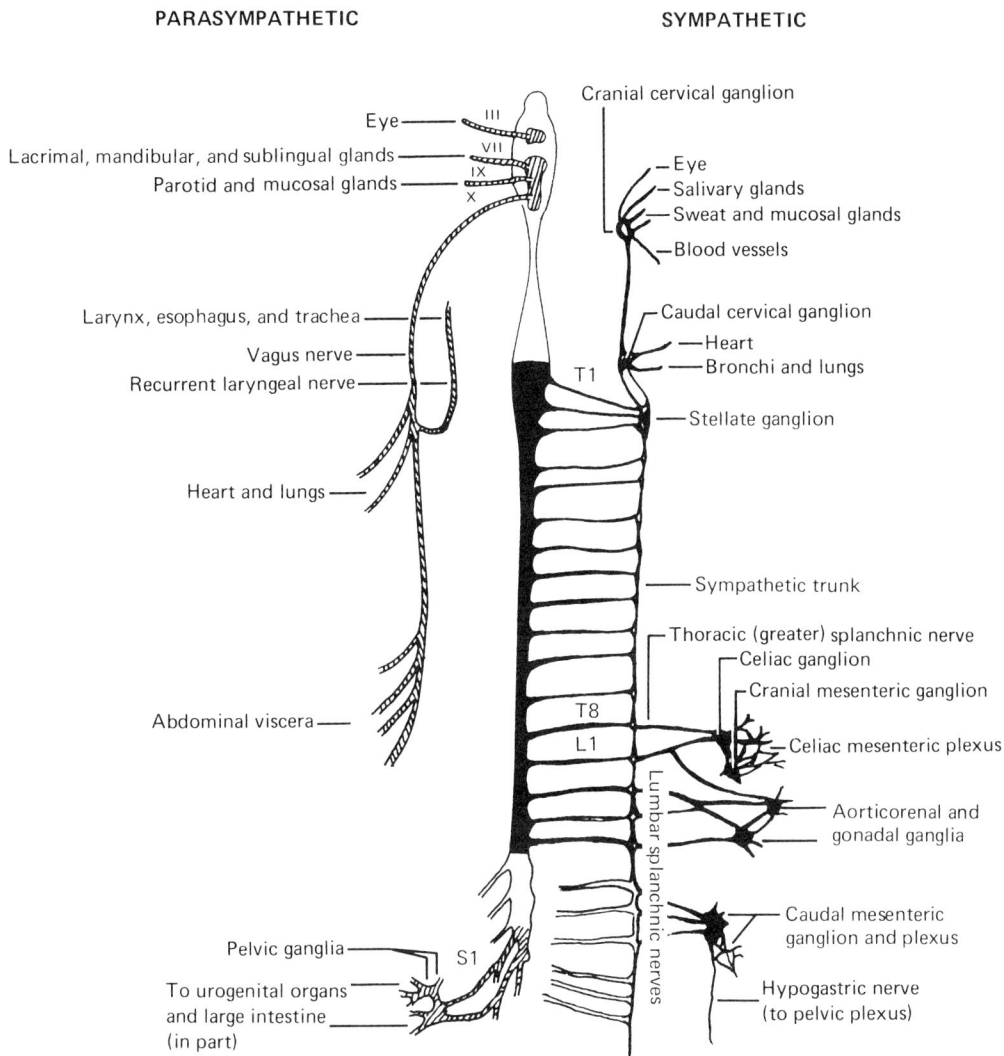

Fig. 8-3. Schematic representation of the autonomic nervous system. (Modified from Miller, M. E., Christensen, G. C., and Evans, H. E.: Anatomy of the dog, Philadelphia, 1964, W. B. Saunders Co.)

the **craniosacral nerves** oppose one another in the effect of their stimuli. The former usually stimulates or increases the action of the parts supplied, while the latter usually relaxes or depresses the action of the organ; however, there are exceptions where the reverse is true. For more detailed information, consult a good textbook on physiology.

SPINAL CORD AND SPINAL NERVES
(Figs. 8-4 to 8-6)

Remove the muscles dorsal to the vertebral column. Then with the bone forceps cut away the neural arches of the vertebrae so as to ex-

pose the spinal cord for its entire length, beginning at the foramen magnum.

Study the size of the cord at different levels. The thickening at the level of the thoracic limbs is called the **cervical enlargement.** Notice how the cord becomes more slender in the **thoracic** region and then broadens again in the **lumbar enlargement.** Find the small filament, the **filum terminale (terminal filament),** in which the cord ends. Make a dorsal view drawing to show the foregoing parts.

Observe the cut end of the cord under the compound microscope. Identify the gray matter, disposed in an H-shaped form in the interior,

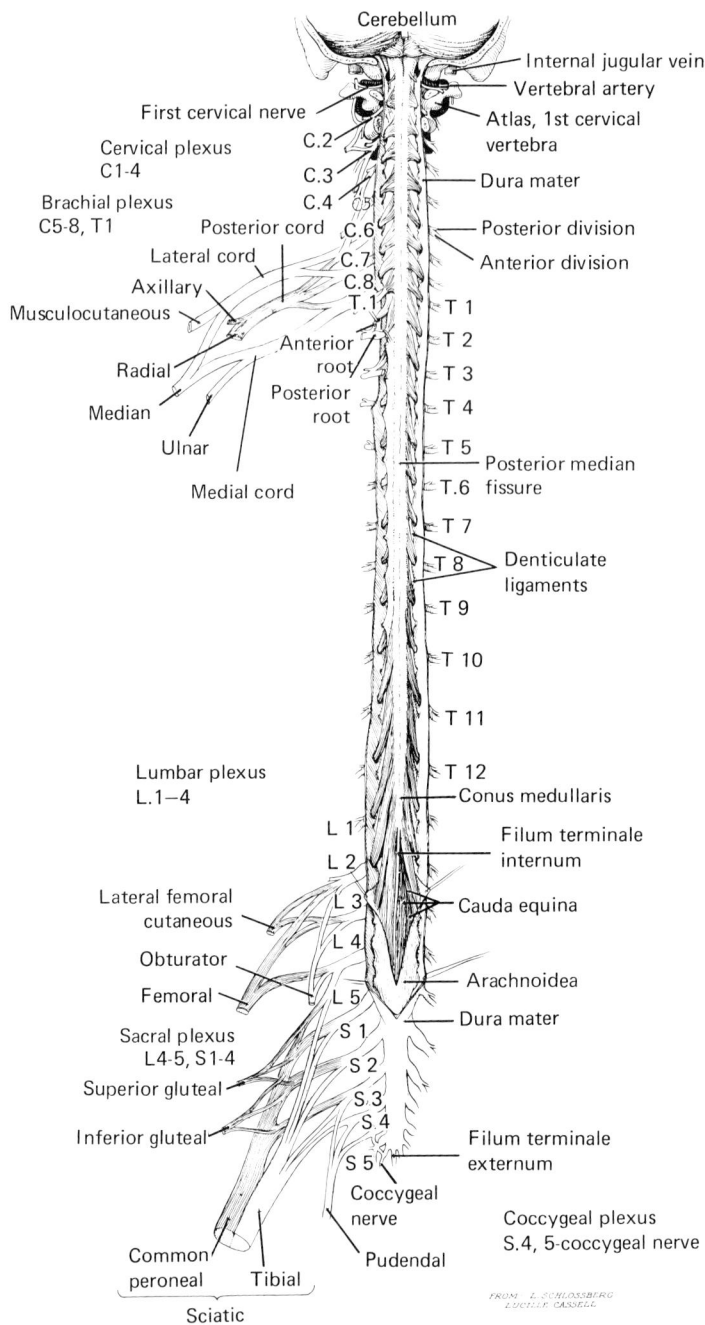

Fig. 8-4. Human spinal cord and spinal nerves. (From Millard, N. D., King, B. G., and Showers, M. J.: Human anatomy and physiology, Philadelphia, 1956, W. B. Saunders Co.)

and the white matter lying outside it. Find the ventral median fissure extending lengthwise along the cord on its ventral side and the median dorsal septum joining the dorsal line.

Remove the neural arches from the sides of the cord sufficiently to expose the origin of the **spinal nerves.** Find that each nerve arises from two roots, **dorsal** and **ventral,** and that these unite with each other to form the nerve a short distance from the cord. Find a prominent rounded swelling on the dorsal root just before it unites with the ventral root; this is the **sensory (dorsal root) ganglion.** Notice the manner in which each of the roots arises from the cord.

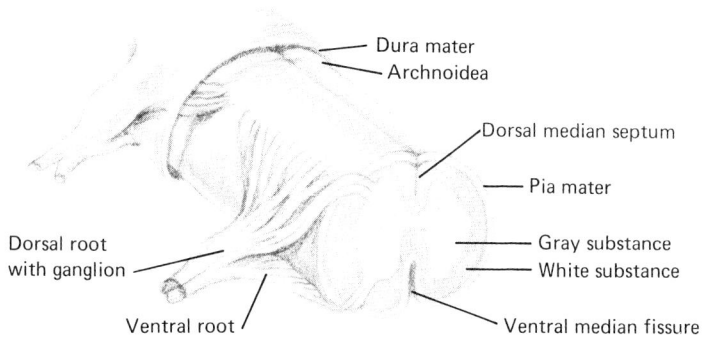

Fig. 8-5. Spinal cord.

Fig. 8-6. Cauda equina.

Note that the nerves in the cranial region of the body leave the cord nearly at right angles and that those in the caudal region come to lie almost parallel to the cord, which is drawn out into a fine, elongated structure and is called the filum terminale. The filum terminale and the lumbar, sacral, and caudal nerves, which lie somewhat parallel and within the neural arches, together constitute the **cauda equina** (Fig. 8-6).

SOME DIFFERENCES IN PLEXUSES AND NERVES OF THE CAT AND MAN

1. In man cervical nerves 5 to 8 and the first thoracic form the brachial plexus, whereas in the cat it is the same except that the fifth cervical is not included.

2. In man cervical nerves 3, 4, and 5 go to form the phrenic plexus, whereas in the cat the fifth and sixth form it.

3. The lumbosacral plexus is formed in the cat by the last four lumbar nerves and three sacral, whereas in man three and one-half lumbar nerves and four sacral nerves form it.

4. There are thirty-eight pairs of spinal nerves in the cat and thirty-one pairs in man. They are arranged as follows:

	Cervical	Thoracic	Lumbar	Sacral	Caudal
Cat	8	13	7	3	7
Man	8	12	5	5	1

REVIEW QUESTIONS ON SPINAL NERVES AND PLEXUSES

1. What are the main divisions of the autonomic nervous system? What activities of the body does it regulate?

2. What are the principal parts of the sympathetic nervous system?

3. Name the three plexuses that supply the limbs of a cat.

4. Name the principal nerves of the brachial plexus.

5. Name the principal nerves of the lumbosacral plexus.

6. What do the bases of the principal nerves of each plexus indicate concerning the number of myomeres that went into the formation of each of the paired limbs?

7. What are the principal parts of the parasympathetic nervous system? Where are they located?

8. Explain about the phrenic nerve and the migration of the diaphragm from its position in the embryo.

9. Name three large ganglia in the neck and thorax.

.

NINE
General summary

This book is intended to present a method of study of the anatomy of the cat and comparisons with man, where the human cadaver is not available for dissection, as a preparation for more exacting courses in human anatomy in medical and dental schools. The main object is not to analyze or interpret the significance of the differences in the anatomy of these two animals; however, several generalizations have been reached, not all of which have convincing proof.

When comparing this carnivore and this primate, we find highly specialized anatomical systems in each. The difference in carriage appears to be the first main factor in effecting the variations and adaptations in all anatomical systems. The cat's body is carried in a more primitive horizontal position, and man's is in an erect position. The animal with the horizontal body position has the advantage of developing more speed than the one with the erect position. The latter, however, has the forelimbs free for defense, for obtaining food, and for taking food to the mouth in suitable amounts and conditions.

The second main factor in the divergence of these two animals is the enlargement and specialization of man's brain. The enlargement forced the skull dorsalward and the nose, jaws, and eyes, ventralward. As a result the eyes could be turned to different horizontal positions by rotating the head or skull on the first vertebra, and likewise the first on the second, without bending the cervical vertebrae from side to side. This ability to rotate the head on the spinal column reduces the need for muscles to move the ears. Therefore the rostral and caudal auricular muscles, which are well developed in the cat's ear, became vestigial in man.

The canine teeth of the cat are fitted for piercing and tearing flesh, not for chewing. These teeth are developed beyond efficient chewing

function in the saber-toothed tiger (saber-toothed cat). Man's canine teeth are relatively small and not well fitted for piercing and tearing, which indicates that no recent ancestors of man followed this method of attack. The movements of the forelimbs of the cat are consistently parallel with the longitudinal vertical plane of the cat's body, and the cat does not need a large clavicle to brace the shoulder as does man, in whom there are movements of the arms from side to side. The coracoid process in both the cat and man is a vestigial remnant of the large coracoid bone of reptiles, whereas in many reptiles and all birds the shoulder lacks a large, flat scapula for the attachment of muscles.

The long neural spines of the thoracic vertebrae, which project caudally (posteriorly), are believed to be a response to the pull of the muscles that support the head on the end of a fairly straight neck. These long neural spines are quite large in the cow and bison, in which the heads are large. The giraffe has an exceptionally long neck but a small head, and the neck is usually held at an angle; therefore, the thoracic neural spines of the giraffe are relatively short. The erect position of man removes much of the pull on the muscles that support the head. The neural spines are relatively small, but their angle suggests that the ancestors of man may have had a horizontal body. In the neural spines of the four caudal thoracic vertebrae, the spines project forward, as in the lumbar vertebrae also, and the transverse processes, because of the pull of the muscles, are from the pelvic region. These long, cranially projecting transverse processes of the lumbar vertebrae indicate that the cat's ancestors were tree climbers or leaping animals. These processes in herbivorous animals and in man project almost straight laterally, showing no signs of recent tree-climbing ancestors.

The sacrum of the cat contains a fusion of three vertebrae, but in man, where there is a greater strain in this region because of the erect position, there are five or seven fused vertebrae. In the bird, where the body is held in a semierect position, the strain is relatively greater, and the sacrum compensates by having a great many vertebrae fused with the pelvis.

There are from four to twenty-six caudal vertebrae in the cat. The long tail with many vertebrae is probably correlated with the ability to run fast. The cheetah, which is noted for its swiftness in running, has an unusually long tail, which is believed to aid in keeping balance when running rapidly. This suggests that the ancestors of the cat were fast runners, and we know that the cat has not lost this ability. The caudal vertebrae of man are reduced to three to five vertebrae that are more or less fused with one another. Sometimes in the human female there are extra vertebrae, or they curve forward, reducing the size of the birth canal, and interfere with the passage of the young at the time of parturition. Cesarean operations are sometimes advised because of this condition. The birth canal is relatively larger in the cat and not so constricted, and parturition is apparently less difficult.

The horizontal position of the body is much better for the support of the intestines, because the dorsal mesentery extends from them to the spinal column. In the erect position the intestines settle or sag down toward the pelvis, constricting or impinging upon the urinary and reproductive systems. Circulative and digestive functions of the intestines are curtailed, and the abdomen tends to protrude.

The erect position in man creates a greater strain on the heart because of the pull of gravity on the blood in most of the veins and on the lymph in the lymphatic vessels. The ankles and calves of the legs often swell in man past middle age, resulting largely from lack of normal lymph circulation. The best rest or relief is obtained when lying down.

The covering of hair on the cat is definitely suited for outdoor life. Man compensates for the lack of hair over the body by wearing clothes, and in cold weather he feels the need of a heavy fur coat. Many of man's anatomical structures and his instinct, however, suggest the need for outdoor life.

The anatomical specializations of the cat are no doubt better adapted for a life in the wilds of nature than those of man for a life in the city. The cat has apparently permitted itself to be partially domesticated because the life is less strenuous. The teeth, the claws, the keen sense of hearing and sight, the strong muscles for swift motion and locomotion, and the ability to digest and live on many kinds of food contribute to the general ability to get along quite successfully without the aid of man. The efficiency of these anatomical structures shows the ability to respond positively to the "call of the wild." The cat has reserve anatomical equipment for offense and defense. The cat shows ability to adapt itself to many conditions of the environment and therefore has strong promise for many future generations.

Man is a highly specialized primate and shows adaptations of many systems, which are not always the best for the kind of life that is now led by a great number of human beings. The greatest specialization in intellectual ability appears at present to be in developing tools of defense and offense and in trips to outer space rather than in developing new methods for the production of more food with less effort. Better methods are needed for the prevention or alleviation of illness and poverty and for the development of better moral standards.

In summary: 1. There are many vestigial organs in cat and man. 2. There are many examples of recapitulation that can be demonstrated. 3. The validity of the existence of vestigial organs and the occurrence of recapitulation has not been seriously questioned. 4. There has not been a more logical or reasonable explanation than the following statements: (a) Vestigial organs are degenerated structures that were more useful and usually relatively larger in lower and possibly ancestral forms. (b) Recapitulation has occurred, continues to occur, and can be demonstrated. From these reasons it is therefore concluded that (1) the cat and man are both highly specialized but in different directions and (2) vestigial organs and recapitulation are interpreted as supporting the belief that the cat and man are what they are structurally largely because of dozens of anatomical changes, which have occurred during the many years of ancestral generations and are largely due to the laws of heredity, response to the environment, and mutation.

Name _____

Date _____

GENERAL REVIEW QUESTIONS

1. Explain and give an example of recapitulation.

2. Explain two anomalies you have observed on the cat.

3. What are the two main parts of the pituitary, or hypophysis?

4. What is the sympathetic nerve chain or trunk? (Fig. 8-3)

5. What anatomic structures of the embryo, or of adult man, suggest a more primitive ancestry?

6. What duct drains each of the following body structures? (a) parotid salivary gland, (b) liver, (c) urinary bladder, and (d) gallbladder. (Figs. 2-10 and 3-2 to 3-4)

7. Do the facts of embryology help in the interpretation of the structure of the adult cat or man?

8. Explain the general function of the hepatic portal system and the liver.

9. Explain the "gorilla" rib in man.

10. State some of the evidence that causes many scientists to believe that man, in his body frame, bears the indelible imprint of his lowly ancestors.

11. State the exact location of the true and the false vocal cords. (Figs. 6-1 and 6-2)

12. How is the ear related to the embryonic gill clefts?

13. Name the two bladders of the cat and the duct that drains each. (Fig. 3-2)

14. From what embryological structures are the cartilages of the larynx and the bones of the middle ear formed?

15. Why is the urethra of the male truly a urogenital duct, while the female urethra is not? (Figs. 3-4 and 3-6)

16. What is a hormone? Where is it produced? (See definitions of terms.)

17. Sketch the branches of the brachiocephalic, carotid, and subclavian arteries as they put off from the aortic arch in the cat and man.

18. How does the ear function in maintaining balance?

19. Examine the hepatic portal system of five cats, and sketch the variations in the way the main branches join one another.

20. Name five glands and their secretions that aid in digestion.

TEN
Definitions of terms

abduction the act of turning outward; a movement away from the midline.

abductor applied to a muscle that draws a structure away from the median line or from a neighboring part or limb.

acetabulum a socket in the os coxae in which the head of the femur articulates.

adduction a movement toward the median plane of the body.

adductor applied to a muscle that draws a structure toward the median axis.

adipose of a fatty nature; fat.

adjacent lying near or close to; contiguous.

adrenal an endocrine gland situated near the kidney.

adrenalin a secretion of the adrenal or suprarenal gland.

adventitia the outer coat of an artery or tubular structure.

afferent to carry to.

ala a winglike structure or process.

alimentary pertaining to digestion or to the digestive canal.

allantois an embryonic diverticulum from the hindgut.

alveolus the socket of a tooth, a sac.

amnion an embryonic sac enclosing the embryo.

analogy similarity in appearance or function without identity.

anastomose a communication or network between blood vessels or nerves.

anatomy the science of the structure of the animal body and the relation of the parts.

anlage the primordium, or first part, of a differentiating part of a structure.

anomaly a marked variation from the normal standard.

antebrachium the forearm; elbow to wrist.

anterior situated in front of or toward the front in man; away from the vertebral column.

aponeurosis heavy fascia, white gristly membrane, serving mainly as an investment for muscle; flat tendon.

appendix an outgrowth or process.

artery a blood vessel that carries blood away from the heart.

articulation place where one solid part rubs against another.

atavistic inheritance of a character from a grandparent or remote ancestor.

atrophied shrunken; having undergone diminution in size.

azygous having no fellow; unpaired.

back behind or toward the rear; the side of the trunk nearest the spinal column.

bilateral symmetry an arrangement of parts or organs in such a manner that the corresponding structures are similar on the two sides.

biology the science that deals with living creatures.

biopsy inspection of the living body or a piece removed therefrom.

bisect to divide or cut into two parts.

brachium the upper arm; shoulder to elbow.

branchial arches the cartilages or bones that lie lateral to the pharynx.

branchial pouches the outpocketings, or diverticulae, of the lateral walls of the embryonic pharynx.

bulb a rounded mass or part; an enlargement.

bulbourethral (Cowper's) gland a secreting organ in the wall of the duct that drains the urinary bladder, near the base of the penis.

bulla a dilated or rounded part of the wall of a cavity.

bursa a sac or saclike cavity.

cadaver the human body after death; a corpse.

calyx a cup-shaped organ or cavity; a recess or pelvis of the kidney.

capillaries the minute blood or lymph vessels.

cartilage gristle, a white elastic substance attached to a bone; a precursor of bone.

catabolism destructive metabolism; a throwing down.

caudad an adverb meaning away from the head, or tailward.

caudal an adjective meaning tailward.

cell theory the belief that all organisms are single cells or organizations of cells.

central of or pertaining to the middle point, lines, or plane; toward the center.

cephalic toward the head.

cervix the constricted area between the vagina and the body of the uterus.

chondrocranium the cartilaginous skull.

chorion the outermost embryonic membrane.

chorion frondosum a part of the chorion that forms the embryonic portion of the placenta.

chyle the milky fluid absorbed by the lacteals of the lymph system in the wall of the intestines.

chyme a thick, grayish liquid mass into which food is converted by gastric juice.

circumflex bend about; curved like a bow.

cochlea the spiral portion of the inner ear that contains the epithelial nerve endings of the sense of hearing.

165

concha shell, as pinna of ear.

connective tissue the cells that bind together and support the various structures of the body.

cornu any hornlike projection.

cranial an adjective pertaining to the head or skull, or in the direction of the head.

crest a projecting ridge.

cyst a sac or bladder that contains a liquid.

deep away from the surface.

diencephalon the second lobe of the brain.

differentiation the process by which a cell or group of cells becomes unlike in structure or function from what it was.

dissect to separate or cut apart for anatomical study.

distal remote, farthest from the center or origin, as opposed to proximal; away from the beginning.

dorsad an adverb meaning the same as dorsal.

dorsal an adjective referring to the back or upper surface as opposed to the ventral or lower surface.

ductus deferens (wolffian duct) the excretory duct of the testis.

ectomy to remove from the body.

efferent outgoing from a center as opposed to afferent.

embryo the fetus in its earlier stages of development, especially early in gestation.

embryology the science that deals with the development of the fetus from the egg.

endochondral developed within cartilage.

endocrine secreting internally into the blood; a substance that modifies metabolism.

endoderm (entoderm) one of the primitive germ layers that gives rise to the embryo; internal layer.

endometrium mucosal lining of the uterus.

endothelial the thin lining of blood or lymph vessels.

enzyme a chemical ferment formed by living cells.

epaxial situated above or upon the axis.

epidermis the outermost, nonvascular layer of the skin.

epigenesis the theory that development starts from a structureless cell.

epiglottis the rostroventral cartilage of the larynx.

epimysium the sheath surrounding a muscle.

epiphysis a dorsal projection from the roof of the diencephalon; a part of a long bone next to the joint.

epithelium a thin layer of cells on the surface of the body.

excretion a substance thrown off from the body.

exocrine secreting outwardly, as opposed to endocrine from a duct.

extension a movement whereby two parts become farther apart or straighten.

external refers to the outer surface as opposed to the central part.

extrinsic having origin or attachment outside of an organ or limb.

facet any small plane surface, as where one bone rubs against another.

fascia a sheet of connective tissue that invests and connects muscles.

fat an oily substance that covers connective tissue.

feces the excrement; the intestinal discharge.

fecundity the ability to produce offspring.

fertilization the fusion of sperm and egg.

fetus the unborn offspring of a viviparous animal.

flexion a movement or bending, as at a joint.

foramen a hole or perforation.

foramen cecum a pit on the base of the tongue where the thyroid gland originated.

foramen ovale an opening between the atria or in the sphenoid bone.

fossa a depression or hollow, usually nonarticular.

front the foremost part in locomotion.

fusiform tapering from the center to both ends.

gamete a mature germ cell.

ganoids a group or subclass of fishes.

ganglion a mass of nerve cells that serves as a center of nervous influence, outside the central nervous system.

gene a unit of hereditary or a germinal factor.

genetics the science that deals with the origin of the characteristics of an individual.

gill cleft same as gill slit; an opening to the outside from the pharynx.

gland an organ that separates any fluid from the blood or produces a secretion.

glottis the opening between the vocal folds.

glycogen a form of carbohydrate or starch stored in the liver.

gonad a reproductive gland, as the ovary or testis.

great or greater omentum a fold of dorsal mesentery attached to the greatest curvature of the stomach.

groin place where the hind limb joins the body.

harderian gland a secreting organ of the eye.

hemal pertaining to the blood.

hepatic pertaining to the liver.

hermaphrodite an animal that has parts of both male and female reproductive organs.

hernia a protrusion of an anatomical structure through an abnormal or normal opening, covered by parietal peritoneum.

histology a discourse on cells and the minute structure of tissues and organs; microscopic anatomy.

homology a similarity in origin and structure.

hormone a chemical substance produced in an organ, absorbed and carried by the blood to some distant organ that it excites to a different action.

hypaxial situated ventral to the body axis.

hypophysis same as pituitary; an endocrinal gland ventral to the brain.

hypothesis a tentative theory; a supposition.

ileum the third division of the small intestine.

ilium one of the pelvic bones.

inferior toward the lower end; away from the head.

infra beneath some structure or position.

inguinal pertaining to the groin.

innominate without a name; nameless.

insertion the place of attachment of a muscle to be moved; the more distal end.

insulin an endocrine secretion of the pancreas.

intermediate between two other structures.

internal refers to the central or deepest part.

intrinsic situated internally; with or pertaining only to one part.

irritability the quality of responding to a stimulus.

Jacobson's organ a vestigial structure in the septum of the nose; the vomeronasal organ.

jejunum the second division of the small intestine.

karyokinesis mitosis; indirect cell division.

lacteals lymph capillaries that absorb fat from the wall of the small intestines.

larynx the voice box or Adam's apple.

lateral to the side of a median plane away from the midline of the body.

leg portion occupied by tibia and fibula.

lesser omentum a part of the ventral mesentery attached to the concave surface of the stomach.

leukocyte a white blood corpuscle.

ligament a form of connective tissue that joins one bone to another.

linea alba the white or median line of the abdomen.

litter the number of young born at one time from the same parent.

loculus a small space or cavity.

lymph a transparent, slightly yellow liquid that arises from the blood and functioning protoplasm.

malleolus a little hammer; a rounded process on either side of the ankle.

mammary pertaining to the milk-producing gland.

marsupium an external pouch in which the young are carried, as in the opossum or kangaroo.

mechanistic physical and chemical forces independent of any life processes.

Meckel's diverticulum a remnant of the yolk stalk attached to the ileum.

medial toward the midline of the body.

median the plane that vertically bisects an organism or part; the midline of the body.

mediastinum a space in the thorax between the two pleurae.

medulla the fifth or most caudal lobe of the brain.

mendelism the law that an offspring is not intermediate in type between its parents but one or the other is predominant.

meninges the membranes covering the brain and spinal cord.

mesoderm the middle primitive embryonic germ layer.

mesonephros the middle kidney or wolffian body.

mesothelium the lining of the body cavity.

metabolism the sum of all processes by which living substance is produced and maintained.

metamere a primitive segment or part.

metamerism a state of being in which the component parts are identical.

metanephros the hind kidney or that of adult mammal.

morphology the science that deals with the form of organized beings.

mucoid resembling mucus.

mucosa the lining of organs that open to the exterior.

mucus a viscid, watery secretion of the mucous glands. (The word "mucus" is a noun, and "mucous" is an adjective.)

myotome that portion of a somite that develops into voluntary muscle; also called myomere.

natural selection a theory devised by Darwin to explain how evolution operates.

navel the scar on the abdomen where the umbilical cord was attached in the embryonic stage.

nephridium a kidney tubule.

nephros a kidney.

neuraxis the brain and spinal cord.

obturator foramen an opening between the ischium and pubic bones in the pelvis.

odontoid process a toothlike projection on the axis.

omentum a double fold of lacelike peritoneum extending from the greater or the lesser curvature of the stomach

ontogeny the development of a single individual organism from the egg.

oogenesis the origin and progressive development of the female germ cell.

organ any part of the body having a special function.

origin the beginning; the end closer to the median plane of the body; the end of the muscle that is the more stationary.

oviparous producing eggs or ova that are hatched outside the body.

parenchyma the essential or functional cells of an organ as distinguished from its stroma or framework.

parietal refers to the body wall.

parthenogenesis the development of an egg or ovum without fertilization.

parturition the act of giving birth to young.

pathology the science that deals with diseased tissue or organs.

pectoral girdle the various bones to which the thoracic limbs or forelimbs are attached.

pelvic girdle the bones, making up the pelvis.

peripheral close to the outer surface; away from the center.

peritoneum the thin, shiny lining of the body wall.

pharynx a section of the digestive tract between the mouth and esophagus; also has respiratory connections.

phrenic related to the diaphragm.

phylogeny the history of a race or a group of plants or animals.

pinna the part of the ear that projects from the side of the head.

pituitary (hypophysis) a gland at the base of the brain.

placenta an organ of pregnancy through which nutriment and waste pass to and from the embryo.

plasma the fluid portion of the blood.

plexus a net or meshwork of nerves or blood vessels.

posterior behind or tailward in animals; dorsal or back in man.

precursor one that precedes; a forerunner.

primordium the first part of a structure to form.

pronation a rotation, as in turning the palm downward or backward.

pronephros the earliest embryonic or head kidney.

prostate gland a secreting organ of the male at the upper end of the urethra.

protraction the pulling forward of a part.

proximal nearest to, relatively closer to the central portion, as opposed to distal; the beginning of a structure.

ramus a primary division of a bone, nerve, or blood vessel; branch.

raphe a ridge, furrow, or line that marks the union of halves of various symmetrical structures.

reflect to turn back.

regeneration the renewal or repair of an injured or absent structure.

regurgitate spit up or vomit.

retraction the act of drawing back.

retroperitoneal behind the thin, shiny lining of the body cavity, as the location of kidneys.

rostral adjective indicating direction opposed to caudal in reference to structures on or within the head.

sagittal resembling an arrow; straight; running vertically parallel to long axis of body.

scrotum a pouch that contains the testes.

sella turcica a Turkish saddle; a depression in the upper surface of the sphenoid bone, the pituitary fossa.

semen the fluid secreted by the testes and other glands that contains the sperm.

seminal vesicle an enlargement or receptacle for sperm near the wreathed end of the vas deferens or sperm duct.

shank the leg between the knee and ankle.

shin the front part of the leg; front of tibia.

sinus a recess; cavity.

somatic pertaining to the framework of the body, as distinguished from the viscera.

somite a primitive segment, a blocklike mass of mesodermal cells in a young embryo.

spermatogenesis the origin and development of the male germ cells.

spheniod wedge shaped; a bone at the base of the skull.

squamosal pertaining to a scale; the vertical plate of the temporal bone.

sternebrae the bony sections of the sternum.

stroma the tissue that forms the framework or matrix of an organ.

superficial pertaining to or near the surface.

superior higher or upper

supination a movement by which palm or plantar surfaces are turned upward or inward.

supinator applied to a muscle that turns the palm upward, as in rotation of forearm.

supra above; beyond.

system an association of organs for the performance of some general function of the body.

symphysis a line of junction or fusion of bones; originally distinct, often of corresponding parts.

tendon a form of connective tissue by which a muscle is attached to a bone.

tentorium a part of the dura mater separating the cerebrum from the cerebellum; sometimes ossifies.

testicle same as testis.

testis the male gonad where the sperm develop.

tissue a group of similar cells united in the performance of a special function.

transverse across; from one side to the other; often means a cross section.

trochanter process of the femur.

tubercle a rough, round eminence on a bone.

tuberosity rough eminence situated on a bone.

tympanum the eardrum.

umbilicus the navel; the scar on the abdomen that marks the place of attachment of the cord to the placenta.

urachus portion of allantois from the urinary bladder to the umbilicus.

ureter the duct that drains urine from the kidney to the urinary bladder.

urethra the duct or passageway from the urinary bladder to the exterior surface of the body.

uterus the womb or place of development of the embryo and fetus.

vagina masculinus (prostatic utricle) the remnant of the müllerian duct found in the male near the upper end of the urethra.

vasa efferentia the tubules that drain the testis.

vein a vessel that conveys blood toward the heart.

ventral pertaining to or situated toward the abdomen; opposite to dorsal.

ventricle any small cavity as applied to the heart or brain.

vermiform wormlike.

vertebra one of the bony sections or units of the spinal column.

vertebrae two or more of the bony sections of the spinal column.

vertebrate an animal that has a spinal column.

vesicle a small bladder or sac containing liquid.

vestigial pertaining to a remnant, rudimentary structure.

villus a minute vascular chorionic tuft.

viscera the large organs of any of the great cavities of the body, especially the abdomen.

viviparous bringing forth young alive.

wolffian duct sperm duct, or vas deferens, extending from the testis to the urethra.

xiphoid shaped like a sword; pertaining to the caudal portion of the sternum.

zygapophysis two or more processes yoked or joined, as in an articular process.

zygomatic pertaining to the arch formed by the malar and temporal bones.